BASIC TECHNICAL DRAWING

PHILLIP SELL

Highline Community College

Merrill, an imprint of
Macmillan Publishing Company
New York

Collier Macmillan Canada, Inc.
Toronto

Maxwell Macmillan International Publishing Group
New York   Oxford   Singapore   Sydney

Editor: Stephen Helba

Developmental Editor: Monica Ohlinger

Production Editor: Constantina Geldis

Cover Designer: Russ Maselli

This book was set in Times Roman.

Macmillan Publishing Company
866 Third Avenue, New York, New York 10022

Collier Macmillan Canada, Inc.

Library of Congress Cataloging-in-Publication Data
Sell, Phillip.
  Basic technical drawing / Phillip Sell.
      p.   cm.
  Includes index.
  ISBN 0-675-21001-1
  1. Mechanical drawing. I. Title.
T353.S412 1991
604.2 – dc20   90-40871
                                                                      CIP

Printing: 1 2 3 4 5 6 7 8 9   Year: 91 92 93 94

MERRILL'S INTERNATIONAL SERIES
IN ENGINEERING TECHNOLOGY

PREFACE

Technical drawings are the means by which design ideas are communicated to the many people who carry out those ideas. Even in this era of computer-aided drafting and design, projects that are originated by architects and engineers become drawings that are prepared by drafters. These drawings are used by planners to plan production, by fabricators to do the actual manufacturing, and by inspectors to make sure the products are made to specifications. Each of these professionals take from the drawings information they need to perform their particular job functions. Consequently, all of these users must understand the language of technical drawing in an easily comprehensible way, which is how the information is presented in *Basic Technical Drawing*.

As the title indicates, this textbook covers the basics of technical drawing. It is written for freshmen engineering technology and technician students who have no background in technical drawing. The emphasis throughout is equally divided between techniques of drawing and understanding the principles of shape and size description. Manufacturing-type drawing is emphasized; however, the principle features of other types of drawings are also covered. All information conforms to current standard practices.

This text is suitable for any beginning, one term drafting or technical drawing course at the college, community college, or vocational school level. It is intended for students with little or no knowledge of the subject who need basic technical drawing or reading skills—either as ends in themselves or as prerequisites for more advanced drafting or graphics courses. It could also be used in an introductory blueprint reading course.

Basic Technical Drawing covers the essential concepts necessary for a fundamental understanding of technical drawings. The topics include freehand sketching, lettering and instrument linework techniques, geometric constructions, pictorial drawing techniques, orthographic projection principles, basic multiview drawing techniques, section views, simple auxiliary views, basic dimensioning practices, and an introduction to working drawings.

The text is designed to be flexible. Topics are arranged so that they may be used in a variety of combinations and sequences. Chapters 2 through 8 are suitable for a beginning drafting or technical drawing course. I prefer, in my own courses, to teach sketching techniques before instrument drawing, so sketching is covered in Chapter 1. For those who do not like this sequence, Chapter 1 is written in such a way that it can be used anywhere. Chapter 9 includes a fairly extensive discussion of working drawings, including the formats and conventions used on production, construction, and schematic types. The emphasis in this chapter is on the understanding of working drawings rather than the creation of them. Chapters 5, 7, 8, and 9 can be used for a basic blueprint reading course.

Features of the text include the following:

- Each chapter opens with an illustrated essay designed to interest and motivate students.

- Chapters begin with behavioral objectives.

- Every chapter ends with a quiz that serves as a summary and review of the chapter's important topics.

- Chapters conclude with extensive problem sets graded in difficulty and selected to provide ample practice with the chapter material.

- The text includes a Glossary of Terms and these important appendixes: Conversion Charts, Tables of Standard Abbreviations, and Tables of Section Line, Electronic, Fluid Power, Pipe Fitting, and Welding Symbols.

The text is accompanied by an Instructor's Manual that includes solutions to all drawing problems and answers to chapter quizzes.

In its preliminary form, the text and many of its problems have been used successfully for over three years in a one-term, five-credit course taught by several different instructors at Highline Community College. Evaluations by faculty and students have been excellent, and I have incorporated their suggestions into the present book.

I would like to thank the following reviewers for their helpful suggestions: J. David Alpert, Augusta Area Technical School; Jim Fitzpatrick, York Technical College; Andrew Ford, Washtenaw Community College; Barry Gornick, Manhattan Technical Institute; David Price, Orange Coast College; Russ Schultz, Hawkeye Institute of Technology; Marc Steiener, Dennco Tech; Jack Swearman, Montgomery College; Alfred Watson, Kilgore College; and James Wiederspan, Western Iowa Technical Community College.

Additionally, I would like to express my appreciation to a number of people for their help in preparing this book. My thanks go to Roger Powell for his initial encouragement of the project; Carol Marquardt for drawing many of the problems; my colleagues at Highline Community College for their many helpful suggestions; my son, Terry Sell, for proofreading and reviewing much of the text; and Monica Ohlinger, my editor at Merrill, who guided me along the winding path of turning a collection of ideas into a book. Most especially, I wish to thank my wife, Sally, and dedicate this book to her for the many hours she spent typing, proofreading, and retyping, and for her never ending patience and love.

CONTENTS

Chapter 6
BASIC MULTIVIEW DRAWING TECHNIQUES   175

Chapter 7
ADVANCED MULTIVIEW TECHNIQUES   215

Chapter 8
DIMENSIONS AND SPECIFICATIONS
249

Chapter 9
WORKING DRAWINGS
283

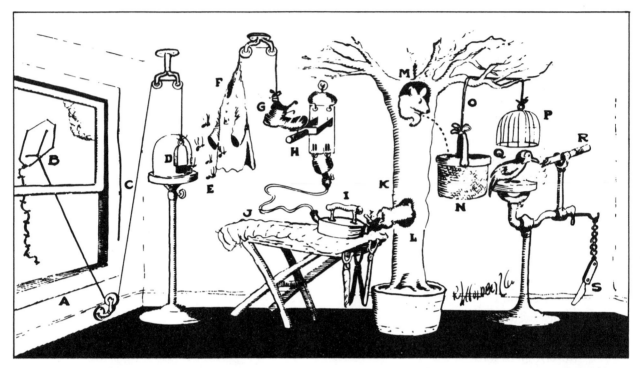

Professor Butts gets his think-tank working and evolves the simplified pencil sharpener.

Open window (**A**) and fly kite (**B**). String (**C**) lifts small door (**D**), allowing moths (**E**) to escape and eat red flannel shirt (**F**). As weight of shirt becomes less, shoe (**G**) steps on switch (**H**) which heats electric iron (**I**) and burns hole in pants (**J**).

Smoke (**K**) enters hole in tree (**L**), smoking out opossum (**M**) which jumps into basket (**N**), pulling rope (**O**) and lifting cage (**P**), allowing woodpecker (**Q**) to chew wood from pencil (**R**), exposing lead. Emergency knife (**S**) is always handy in case opossum or the woodpecker gets sick and can't work.

1

FREEHAND DRAWING
AND LETTERING

1-1 INTRODUCTION

"A picture is worth a thousand words" is an old saying that means it is easier
to describe an object by drawing a picture of it than by describing it in words.
Choose any simple object and think about how difficult it would be to describe it accurately to someone by using verbal communication only. Drawing
objects is the most effective way to show others exactly what the objects are.

The illustration that opens this chapter is a Rube Goldberg cartoon showing an automatic "pencil sharpener." You aren't supposed to take this invention
seriously, but to understand the joke, you must be able to "read" the drawing
and follow the explanation. This cartoon is a good example of the use of freehand drawing and lettering. Though this chapter is not about cartooning, it
is about the same basic skills and techniques that cartoonists use. These are
the same skills that you may find useful in your everyday life. Have you ever
tried to draw a map to your house? A floor plan of your apartment? Or tried
to describe the gizmo that broke on your car? Skill in creating quick, clear,
and accurate freehand sketches would certainly make these types of commu-
nications better and easier.

In this chapter you will study sketching and lettering, which are freehand
methods of creating technical drawings. These topics are covered first
because they are the simplest and quickest methods for communicating technical information. This chapter will also introduce multiviews and pictorials,
the two principal types of technical drawings.

After you complete this chapter,
you should:

1. know basic freehand sketching techniques
2. know the principal types of
lines used in technical
drawing
3. know the essential features
of two-dimensional (multiview) drawings
4. know the essential features
of three-dimensional (isometric) drawings
5. be able to do clear, accurate
freehand drawings of simple
objects

1

1-2 FREEHAND DRAWING

One form of drawing that is widely used by many people—but especially by engineers, architects, and drafters—is the freehand sketch. It is useful for quickly recording ideas and designs, for working out preliminary problem solutions, and as an aid in explaining concepts to others.

Sketching Techniques

Sketches may be drawn entirely freehand or with the aid of drawing instruments. Using instruments, such as a straightedge, will aid you in producing regular line work but will also slow you down. With practice you will be able to make a quick, but very acceptable, drawing entirely freehand. Paper, pencils, and eraser are the only items required. Any type of paper can be used, but paper with grid lines is usually the easiest to work with. Use a medium-grade (H or F) or No. 2 pencil, and keep the point medium sharp. Any type of soft eraser will do.

When sketching freehand, rest your forearm on the table and draw lines in lengths corresponding to the natural movement of your hand and arm. Use short overlapping strokes with your pencil. Keep your eye on the point you are drawing to, not on the line you are drawing. Sketched lines are not supposed to be perfectly uniform. Do not tape your paper to the drawing surface. This will allow you to draw lines in the direction that is most comfortable for you by turning the paper as necessary. Don't be afraid to erase and redraw lines as needed.

Sketching on graph paper helps you achieve uniform proportions in your drawing. Sketches are seldom drawn to an exact size, but for maximum usefulness they must be drawn proportionally. For example, if the object is 2 inches long and 1 inch wide, it is seldom necessary to draw your sketch to those exact dimensions. But the object should be drawn approximately twice as long as it is wide.

Two general types of lines, straight lines and curved lines, are used in technical drawing. **Straight lines** are any lines that continue in the same direction for their entire length. Straight lines are often identified by their direction. **Vertical** lines have an up-and-down direction. **Horizontal** lines have a level or side-to-side direction. **Inclined** lines have a slanting direction—that is, any direction not vertical or horizontal. See Figure 1-1 for examples of lines drawn in different directions.

All straight lines drawn on paper are really horizontal lines as long as the paper is lying on a horizontal surface. But technical drawing uses a convention that says that any line drawn from the top toward the bottom of the paper

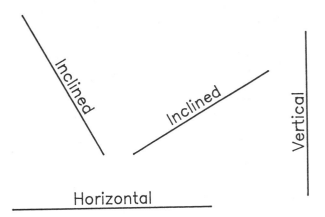

Figure 1-1 Vertical, horizontal, and inclined straight lines.

is a vertical line and any line drawn from one side to the other is a horizontal line. Any other straight line is called an inclined line.

Two or more straight lines that are drawn in the same direction—that is, they stay the same distance apart—are called **parallel lines.** Figure 1-2 shows three sets of parallel lines. Note that parallel lines may be drawn in any direction.

Two straight lines drawn in exactly opposite directions (for example, one horizontal and one vertical) are called **perpendicular lines.** Figure 1-3 shows several examples. Each set of perpendicular lines forms a right (90°) angle. Note that perpendicular lines may be drawn in any direction.

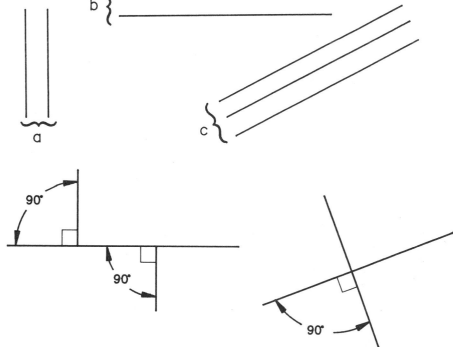

Figure 1-2 Three sets of parallel lines: (a) vertical, (b) horizontal, and (c) inclined.

Figure 1-3 Perpendicular lines. The small square shown at each intersection indicates perpendicularity.

Since there are only two general types of lines, straight and curved, any line that is not straight is a **curved** line. A curved line's direction is constantly changing, and the way it changes determines whether the line is a definable curve, such as a circle, or an irregular curve. Circles are the most important curved lines in technical drawing. **Circles** are lines that begin and end at the same point, and are everywhere the same distance from a point called the center. This distance is called the **radius** (abbreviated R), and twice the radius is called the **diameter** (abbreviated Ø or D). A portion of a circle is called an **arc. Irregular curves,** unlike circles, have no defining points, such as centers, or dimensions, such as diameters. Each irregular curve is unique. Circles, arcs, and irregular curves are illustrated in Figure 1-4. Note that the circles are divided into four equal parts by broken lines that intersect at the center. This type of line is called a **centerline** (abbreviated ₵). Centerlines are used to indicate that the circles are **symmetrical**. This means they are exactly the same (or a mirror image) on both sides of each centerline.

Somewhat different techniques are required for sketching different types of lines. Figure 1-5 shows the best positions for your hand and pencil for sketching horizontal, vertical, and inclined lines. All straight lines are best sketched using total arm movement. Keep your wrist stiff and move primarily with your shoulder and elbow. Horizontal lines are most easily drawn from

Figure 1-4 (a) circles, (b) arcs, and (c) irregular curves.

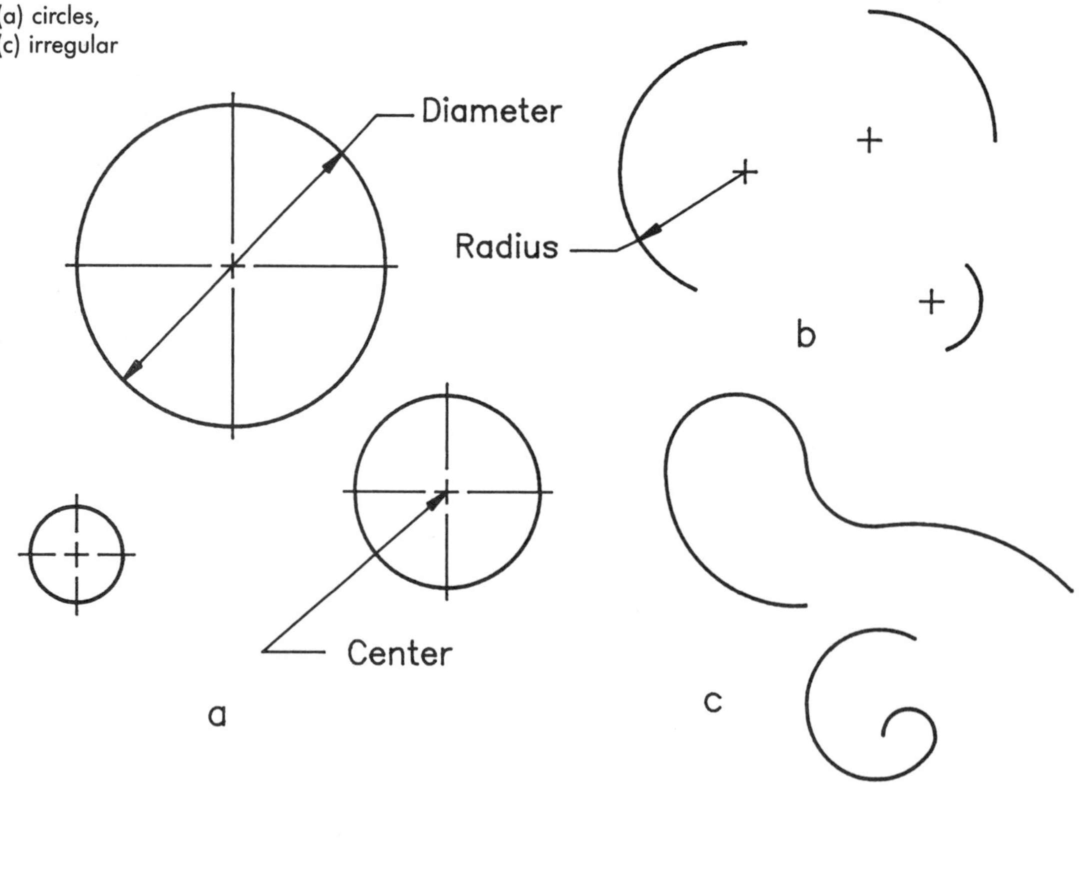

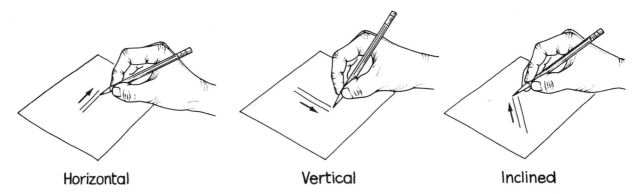

Horizontal   Vertical   Inclined

left to right, if you are right-handed, or right to left, if you are left-handed. Vertical lines are most easily drawn from top to bottom. The ease of drawing slanted lines varies with their direction. However, if you rotate your paper as previously suggested, all lines can be drawn in whatever direction is most comfortable for you.

Figure 1-6 shows examples of correctly sketched straight lines. Horizontal, vertical, and inclined lines are shown in Figure 1-6a. Notice that the inter-

Figure 1-5 Hand and pencil positions for drawing horizontal, vertical, and inclined lines.

Figure 1-6 Correctly sketched (a) solid lines, (b) dashed lines, and (c) centerlines.

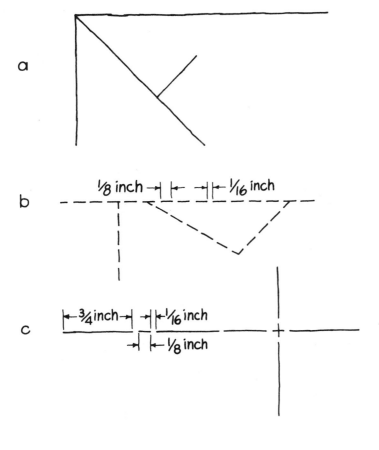

sections, or points where the lines meet, are sharp and clean with little over-lap. Clean intersections help make your drawing clear and are best achieved by starting at the intersection with your pencil and drawing away from it. As Figure 1-6b shows, dashed lines are used to indicate, among other things, hidden features. The dashes should be about 3 mm to 6 mm or .10 in. to .25 in. long. Sketched lines won't have the same precision as those drawn using instruments.

There is no preferred hand position or direction for sketching curved lines. Use whatever is most comfortable for you. Unlike sketching straight lines, it may be easiest to sketch circles or arcs with wrist movement.

Circles that look like circles are very difficult to draw freehand. When sketched in a hurry, they often turn out looking like anything but circles. Figure 1-7a shows techniques for sketching reasonably accurate circles. Start by sketching two light perpendicular, intersecting lines and marking a radius of the circle on each line. Draw diagonal lines that pass through the center, and mark radii on these also. Mark as many radii as you need to establish the shape. Connect all the marks, and you draw a reasonable-looking circle. As

Figure 1-7 (a) The steps in sketching a circle. Construction lines may be erased after step 5, if they are too dark. (b) Examples of sketched arcs.

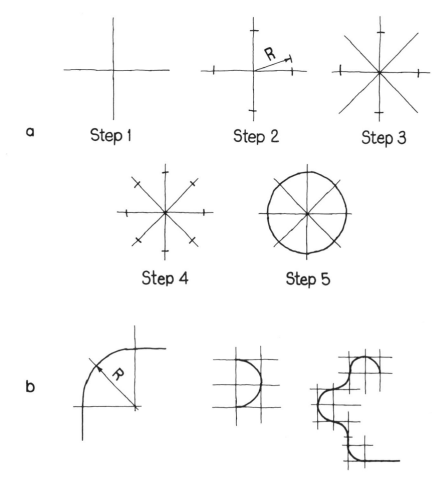

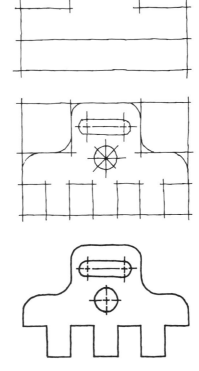

Figure 1-8 The steps in making a freehand sketch.

Step 1. Sketch major features with light construction lines.

Step 2. Sketch details with light construction lines.

Step 3. Finish by darkening lines and erasing construction lines.

Figure 1-7b shows, you can use the same method to draw arcs, which are, after all, nothing more than portions of circles.

As Figure 1-8 shows, drawing a complete sketch of an object is a three-step process. First, sketch the major features and centerlines with light temporary lines called **construction lines.** Pay particular attention to the proportions of the object you are drawing. Second, sketch in all the details, including circles and arcs, again with light construction lines. Third, finish your drawing by erasing construction lines that are too dark and darkening those lines you want to keep. The technique of drawing light construction lines first and then darkening those that define the object will be emphasized throughout this text. This method of constructing a drawing by first putting in the major features and outlines and then filling in the details is called **box construction.**

Pictorial sketching. Look around the room you are in. Select some object in a far corner and study it briefly. Ask yourself this question: If you wanted to describe that object by drawing a picture of it, would you choose the view you see from where you are now sitting, or would you move to a different place to gain a better view? Choosing the view or views to draw is very important in technical drawing because you are always trying to communicate information as clearly and accurately as possible. Views of objects taken from whatever position you may happen to be in will seldom do the job. Some views describe objects better than others. Technical drawing includes two standard ways of viewing objects: pictorial drawing and multiview drawing.

Pictorial drawings show objects in approximately the same way photographs do. They are attempts to create three-dimensional pictures with relatively realistic appearances. The discussion in this chapter will be limited to a type of pictorial drawing called isometric drawing, but there are several other types of pictorial drawing that are covered in Chapter 4. In an **isometric drawing** the object is oriented so that one of its corners (if it has any) is closer to you than the rest. As Figure 1-9 shows, an isometric drawing creates the illusion of receding lines and a feeling of three-dimensional depth.

The easiest way to create an isometric drawing is to start by drawing a box that is just big enough to contain your object. This is called an isometric box, and it should be drawn so that its receding lines are 30° to a horizontal line and its front edge is vertical (see Figure 1-10).

Determine the proportions of the object to be drawn, and start sketching the details. Figure 1-11 shows how to begin defining a shape within an isometric box. Continue until all the visible features are included. When you are satisfied with your drawing, complete it by darkening and erasing lines as needed.

Figure 1-12 shows several examples of isometric sketches. The isometric boxes are still visible so you can see how they were drawn.

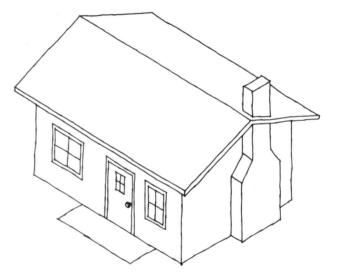

Figure 1-9   An isometric sketch.

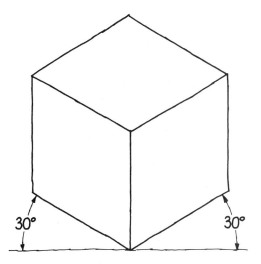

Figure 1-10 The isometric box is the best starting point for most isometric sketches.

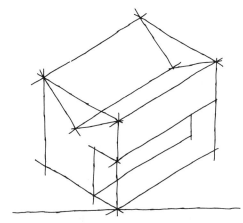
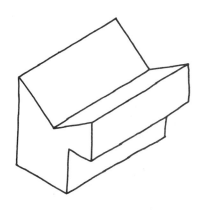

Figure 1-11 Completing an isometric sketch.

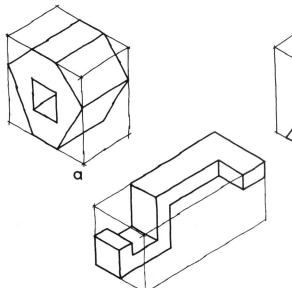
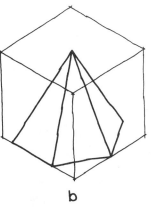

Figure 1-12 Examples of isometric sketches.

Figure 1-13 (a) Circles appear as ellipses in an isometric drawing. (b) An isometric cylinder. (c) Elliptical arcs.

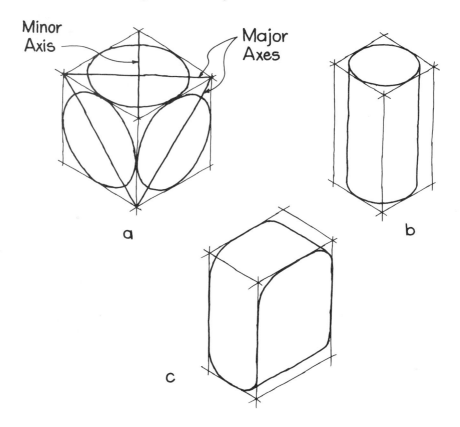

In an isometric drawing a circle always appears as an ellipse. An **ellipse** may be thought of as a tilted circle. A circle has one diameter; an ellipse has two perpendicular centerlines of different lengths. These are called the major axis and the minor axis. Figure 1-13a shows several ellipses drawn on the sides of an isometric box. Notice the location of the major and minor axes. The **major axis** indicates the longest dimension of an ellipse; the **minor axis** indicates the shortest dimension. Cylinders and arcs, as well as ellipses, may be sketched by using the isometric box as a guide. Parts b and c of Figure 1-13 show a cylinder and various arcs drawn as ellipses.

This has been a very brief introduction to pictorial drawing—an introduction intended only to get you started with sketching freehand drawings. Chapter 4 contains much more information on isometric drawing and other types of pictorial drawing.

Multiview Sketching

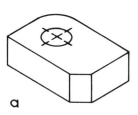

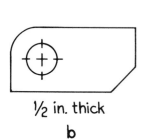

Figure 1-14 A flat object shown (a) as an isometric drawing and (b) as a flat two-dimensional view.

Figure 1-14 shows a flat object with a hole in it. The object is drawn two different ways. In Figure 1-14a, the object is drawn in an isometric view; in Figure 1-14b, it is drawn as a flat two-dimensional view with the thickness indicated by a note. Which view do you suppose is easier to draw? There is

Figure 1-15 A multiview drawing that shows the thickness in a side view.

probably little question that part b is easier to draw but every bit as informative as part a. In a **multiview drawing,** each view is a flat two-dimensional picture showing only one side of the object as in Figure 1-14b. Sometimes it is necessary to present more than one view of a complex object to give all the needed information.

A multiview drawing of the object shown in Figure 1-14 is shown in Figure 1-15. The thickness is shown in the second view, rather than described in a note. Notice that dashed lines are used to indicate the hole that is hidden in the second view.

The positions of views in a multiview drawing are very important to the readability of the drawing. In Figure 1-16e, which shows a completed multiview drawing, notice that the top view is in line with and above the front

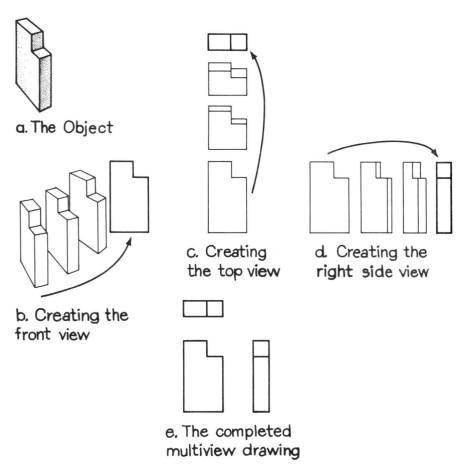

a. The Object

b. Creating the front view

c. Creating the top view

d. Creating the right side view

e. The completed multiview drawing

Figure 1-16 The placement of the views in a multiview drawing reflects the position from which the object is viewed.

Figure 1-17 A one-view drawing, a floor plan of a house.

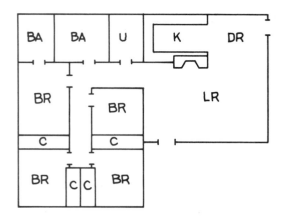

view, and the right side view is in line with, and to the right of the front view. The placement of the views reflects the position from which the object is viewed.

For some drawings one view is all that is necessary. For example, the house floor plan shown in Figure 1-17 requires only one view: the top view. Complex objects, such as the vise shown in Figure 1-18, may require three or more views for a complete description. Notice that Figure 1-18 includes top, front, and right side views.

Figure 1-19 shows three views of a house. The three views shown are the top, front, and left side. Notice their placement. As in previous examples, the top view is drawn above the front view but the left side view is drawn to the left of the front view. The left side view was drawn because it shows more information than the right side view would. Of course, both sides could be drawn if needed.

Figure 1-18 A complex object, such as a vise, needs several views to describe it.

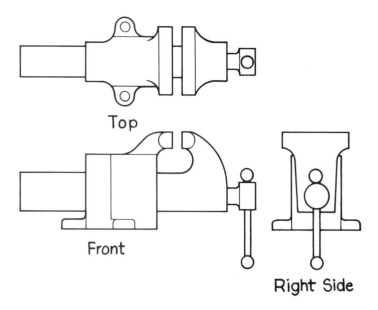

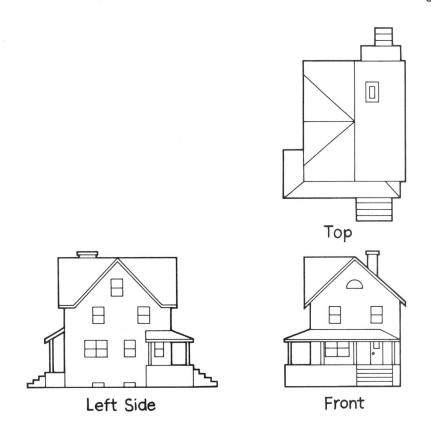

Top

Left Side   Front

Figure 1-19   Three views of a house.

Some objects, such as cars and houses, have natural top, front, and side views. The most effective views of other objects, such as tables and suitcases, are not so readily identifiable. You know which is the top view of a table, but which is the front view? In these cases you must decide which view to name the front and which to name the right side and so on. Whatever views you decide on, it is most important that they be the views that best describe the object. Figure 1-20 shows examples of several multiview drawings. You are not limited to three views, nor are you restricted to always drawing top, front, and side views. You can draw any views that give a clear description.

In this chapter, only simple two- and three-view sketches have been introduced. Multiview drawing of more complex objects is covered in Chapters 5, 6, and 7.

You must decide which type of sketch to make, pictorial or multiview, and which views to draw every time you sketch something. The question that always needs answering before you start sketching is Which type of drawing will most clearly describe the object?

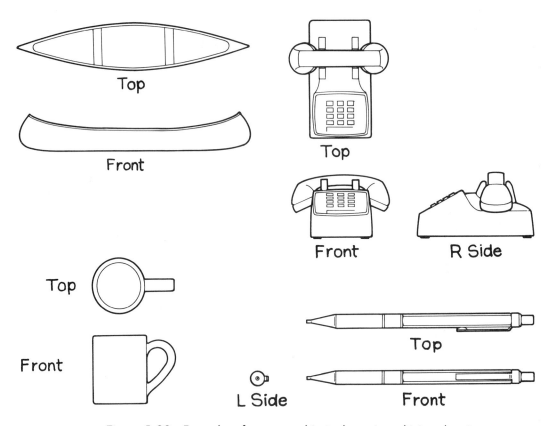

Figure 1-20 Examples of common objects shown in multiview drawings.

1-3 LETTERING

Though pictures are better than words for describing many things, they are seldom sufficient by themselves. A drawing nearly always needs written information to be considered a complete description. Notes, dimensions, and titles are an integral part of all technical drawings. Their clearness is equally as important as the clearness of the picture. Clearness is achieved through the use of properly formed standardized letters and numerals. Figure 1-21 and Figure 1-22 show the shape and style of the simple block letters and numerals commonly used on technical drawings. Note that they may be either vertical or slanted, but, whichever form is chosen, it must be used consistently.

Usually, all letters used in technical drawings are capital letters. The heights should conform to the dimensions listed in Table 1-1. (Sheet sizes are explained in Chapter 2.) Most lettering is 3.5 mm or .125 in.

Figure 1-21 Stroke sequences and proportions for standard letters and numerals.

Figure 1-22 Standard (a) vertical letters and numerals and (b) inclined letters and numerals.

Table 1-1 Recommended Letter Heights

Application	Sheet Size	Approximate Letter Height*
Notes and dimensions	A, B, A3, A4	3.5 mm or .125 in.
	C, D, E, A0, A1, A2	5 mm or .15 in.
Titles, drawing numbers, and view indicators	All sizes	7 mm or .25 in.

* Measurements cited are alternatives, not equivalents.

You can best produce legible, dark letters by using a medium-grade (H or F) or No. 2 pencil. Use single strokes in forming your letters. These are made with a finger and wrist motion, with your forearm resting on the drawing table. The sequence of strokes is generally from top to bottom and from left to right. A definite stroking sequence, such as that shown in Figure 1-21, may be of help. To maintain consistent line width, you must keep your lead medium sharp and rotate it frequently as you letter. Smudges can be eliminated by placing a piece of paper under your hand, as shown in Figure 1-23.

In combining letters to form words, do *not* space the letters equally. Place letters so the space between them *seems* to be equal. Space words so they are as far apart as the width of the letter *O*.

To achieve uniform letter height, use a pair of horizontal parallel guidelines drawn very lightly with a hard (6H) lead. As Figure 1-24a shows, the letters and numerals should touch, but not go past, the upper and lower guidelines. If you have trouble keeping your letters uniformly vertical or slanted, try using vertical or slanted guidelines, as shown in Figure 1-24b. Guidelines may be drawn by using any straightedge or by using special lettering aids like those shown in Figure 1-25. The Ames lettering guide, shown in Figure 1-25a, is used to draw guidelines; the guide shown in Figure 1-25b can be used to draw guidelines or to letter directly within the slots.

Figure 1-23 Use scratch paper to prevent smudges.

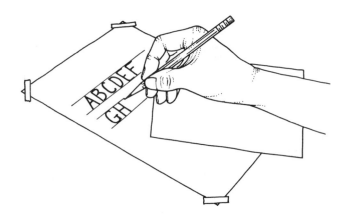

NAME OF SCHOOL : HIGHLINE	
LOCATION: *DES MOINES*	

DRAWN BY: R. MAPLESTONE	2-17
CHECKED BY: R. POWELL	16

a

Figure 1-24 Examples of (a) letter and word spacing and (b) the use of horizontal and vertical guidelines.

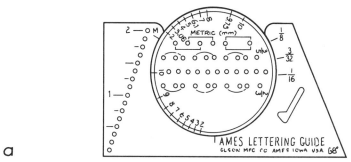

b

Figure 1-25 Two types of lettering aides: (a) the Ames lettering guide and (b) a freehand lettering guide.

a

b

For most people it takes patience and practice to learn the art of lettering. There are no shortcuts. You must follow the procedures given in this section and take your time. You can't rush and achieve good results. Speed will come with practice and experience.

1-4 CONCLUSION

This chapter has introduced technical drawing through freehand sketching and lettering. Sketching drawings is a quick method for drawing objects in an approximate fashion. You cannot achieve great accuracy by drawing freehand, but you can communicate general ideas very efficiently and effectively.

Most of the assignments in this text will require you to use instruments, but don't overlook the value of sketching as a quick way to plan your instrument drawings. It will help you to eliminate mistakes; create a more understandable drawing; and, believe it or not, save you time.

All drawings, freehand or instrument, contain notes and dimensions and thus require lettering. For many, consistent, legible lettering is the most difficult aspect of technical drawing. Unless you have a natural flair for it, the only way to achieve good lettering is practice, practice, practice.

Check your understanding of this chapter by answering the following questions.

REVIEW QUESTIONS

1. Sketched lines are best drawn:

   a. by using short pencil strokes

   b. by using a very hard lead

   c. with single pencil strokes

   d. with the paper taped to the drawing surface

2. Which of the following is required to make sketches?

   a. paper

   b. medium-grade pencil

   c. eraser

   d. all the above

3. Sketches should be drawn:

   a. accurately

   b. proportionally

   c. disproportionally

   d. precisely

4. Two straight lines drawn in the same direction are:

    a. parallel

    b. perpendicular

    c. intersecting

    d. inclined

5. Two lines that form a 90° angle are called:

    a. parallel

    b. perpendicular

    c. intersecting

    d. inclined

6. A horizontal line is:

    a. a slanting line

    b. an oblique line

    c. a level line

    d. none of the above

7. Drawings that attempt to show objects in three-dimensional views are called:

    a. multiviews

    b. projections

    c. sketches

    d. pictorials

8. An isometric sketch is easiest to draw inside a box whose bottom edges are at an angle of _____ in regard to a horizontal line.

    a. 10°

    b. 20°

    c. 30°

    d. 45°

9. Multiview drawings may have as few as:

    a. one view

    b. two views

    c. three views

    d. four views

10. Three standard views most often used in a multiview are:

    a. top, front, and left side

    b. top, front, and right side

    c. top, rear, and right side

    d. bottom, front, and left side

11. Lettering done in technical drawing

  a. must be vertical

  b. can be any form

  c. can be any size

  d. none of the above

12. Lettering guidelines should always be

  a. light parallel lines

  b. dark parallel lines

  c. light dashed lines

  d. dark perpendicular lines

PROBLEMS

1-1 Using technical drawing–style lettering, compose the best verbal description you can of any of the following. Limit your description of each to one page of .125-in.-high lettering.

  a. your pencil

  b. your chair or stool

  c. someone's eyeglasses

  d. any of the objects shown in Figure 1-20

1-2 Make an information sheet for your instructor by lettering the following information in 3.5 mm or .125-in. letters. Follow the format given.

Name:
Course Name and No:
Phone
  Work:
  Home:
Major:
No. of credits completed:
Previous technical drawing courses or experiences:
Reason for taking this course:

1-3 Copy the following using technical drawing–style capital lettering within 3.5-mm or .125-in. guidelines.

GOOD LETTERING TAKES PRACTICE. ONE SHOULD USE A MEDIUM (H OR F) LEAD AND FORM LETTERS AND NUMERALS WITH A SINGLE STROKE. GUIDELINES, ESPECIALLY HORIZONTAL ONES, ARE NECESSARY FOR A GOOD JOB UNLESS SOME TYPE OF LETTERING GUIDE IS USED. OF COURSE, STAYING WITHIN THE GUIDELINES IS ESSENTIAL. MOST LETTERING SHOULD BE 3.5 mm OR .125 in. HIGH – EXCEPT FOR TITLES, WHICH MAY BE AS LARGE AS 7 mm OR .25 in. LEGIBILITY DEPENDS ON CONSISTENCY IN LETTER SIZE AND SHAPE.

1-4 Copy the following sentences within 3.5-mm guidelines. Make your letters look as much as possible like those shown in Figure 1-22.

a. 107 QUICK BROWN FOXES JUMPED OVER THE LAZY DOG.

b. PICKING JUST SIX QUINCES IN 22 DAYS, THE NEW FARMHAND PROVED STRONG BUT LAZY.

c. WE COULD JEOPARDIZE SIX OF THE GUNBOATS BY TWO QUICK MOVES.

d. JUDGE POWER QUICKLY GAVE SIX EMBEZZLERS STIFF SEN-TENCES OF 39 YEARS EACH.

e. THE LAZY JUDGE WAS VERY QUICK TO PAY THE $543.28 IN TAX MONEY FOR THE BARN.

f. JACK TYPED REQUISITIONS FOR 656 WHITE MOVING BOXES (LONG SIZE).

g. I QUICKLY EXPLAINED THAT MANY BIG JOBS INVOLVE FEW HAZARDS.

h. QUICKLY PUT FIVE DOZEN MODERN JUGS IN EACH OF THE 89 BOXES.

i. DAVE QUICKLY SPOTTED THE FOUR WOMEN DOZING IN THE JURY BOX AT 7:00 AM.

1-5 Sketch any of the following:

a. any of the objects you described in Problem 1, part a, b, or c.

b. the top of your table – include instruments, papers, books, etc.

c. your hand

d. your view of the classroom; omit the people

e. a map from your house to school; include names of major streets and dis-tances

f. the floor plan of your house or apartment

g. your car, motorcycle, or bicycle

1-6 Make isometric sketches of these objects at twice the size they are shown.

a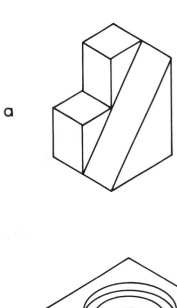

b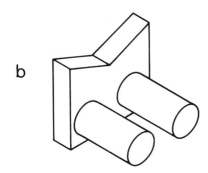

c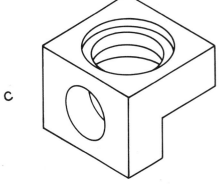

d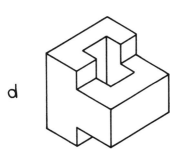

e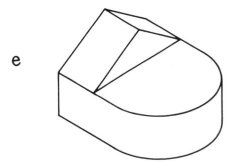

f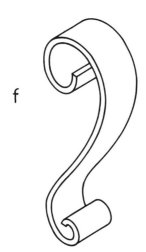

1-7   Make multiview sketches of these objects at twice the size they are shown.

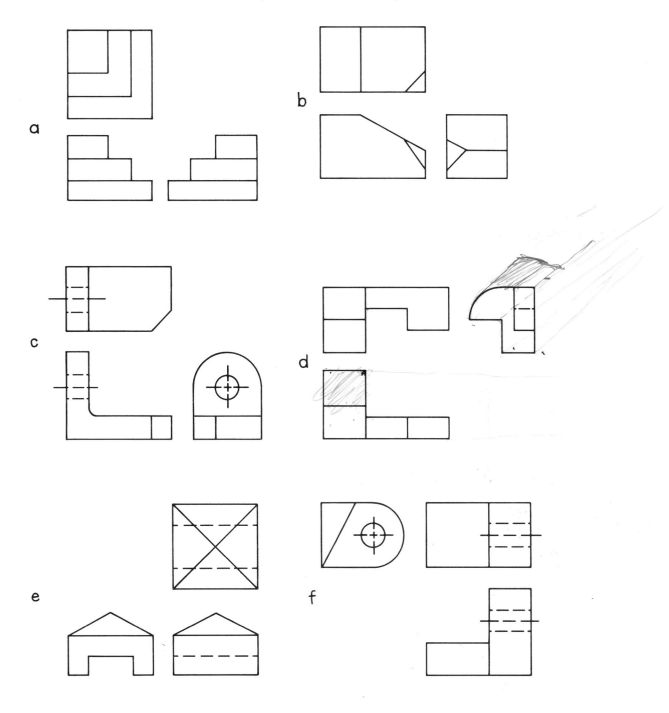

1-8 Make multiview sketches of the objects shown. Show two or three views as required.

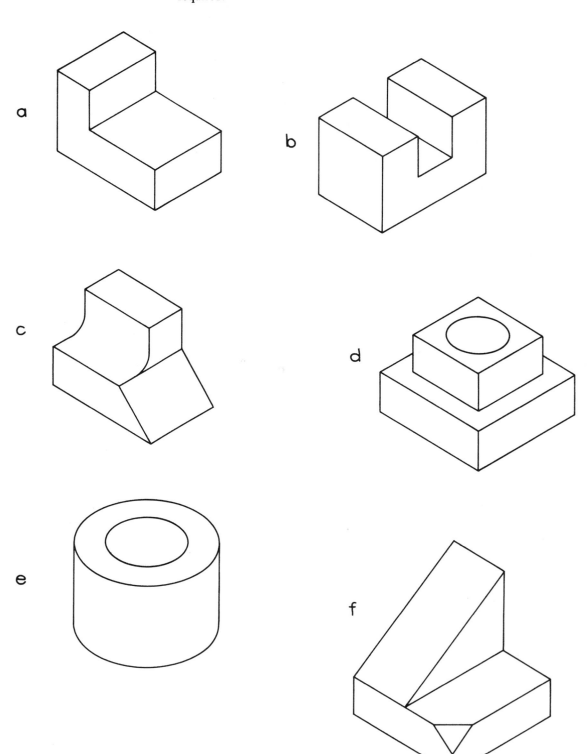

1-9   Make isometric sketches of the objects shown. Orient each object to show the most information.

a
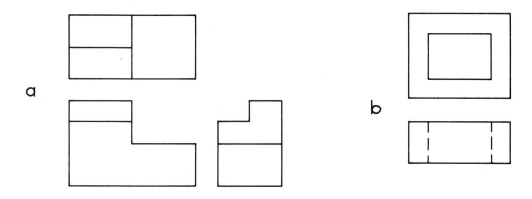

b
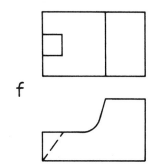

c
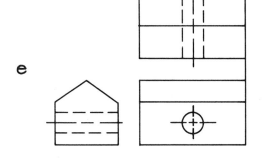

d
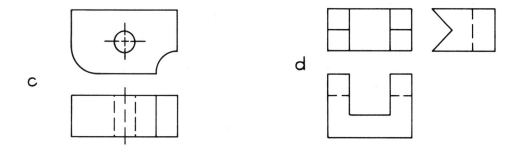

e

f

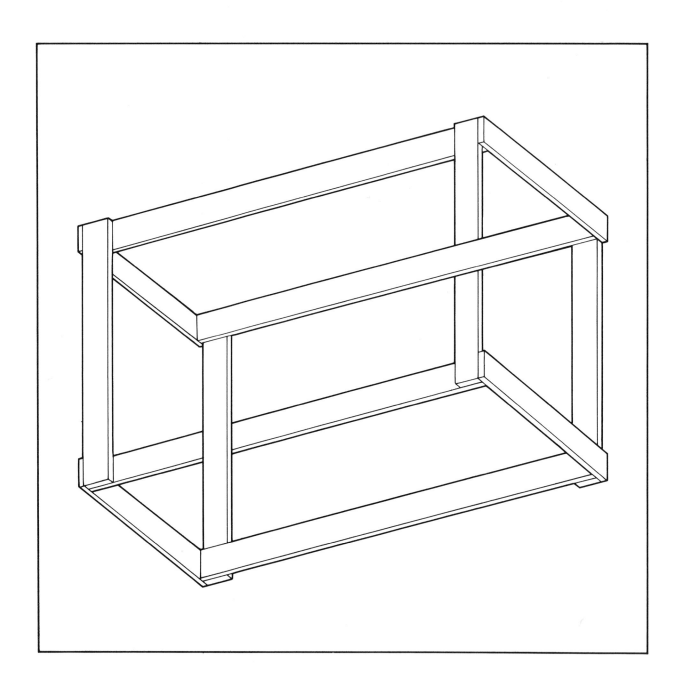

2

LINE WORK AND THE
USE OF INSTRUMENTS

2-1 INTRODUCTION

The purpose of technical drawing is to communicate technical information. To ensure effective communication, technical drawings must be clear and accurate. Clearness, or understandability, is achieved by following certain rules. Writers must use the correct rules of grammar and spelling, and readers must understand the rules in order for communication to take place. Those who use and create technical drawings must understand and follow the rules, or conventions, of technical drawing. Accuracy is achieved through depicting objects as they really are, with neither distortions nor optical illusions.

The illustration that opens this chapter is a drawing that doesn't follow the rules. The result, as you can see, is a picture of an object that can't exist, an optical illusion. Illusions are fun to draw, but such drawings don't communicate useful information because they purposely violate some of the rules. By studying the topics in this book, you can find out which rules have been violated to create this illusion and learn how to draw some illusions of your own.

Clear and accurate drawings are created, in part, by using appropriate drawing materials and instruments and adhering to standards. This chapter describes the most commonly used drawing materials, the fundamental instruments needed for doing basic technical drawing, and the standard line types used in technical drawing. The principal materials include paper, leads, inks, and erasers. Drawing instruments are used to draw smooth regular lines, both straight and curved, and to measure those lines. Each of the standard line types has a particular form and a set of specific meanings. Proper use of the different lines is essential to the understandability and accuracy of technical drawings.

After you complete this chapter, you should:

1. know about the basic types of materials and equipment used in technical drawing

2. be able to use the basic materials and equipment to draw both straight and curved uniform lines

3. be able to use both inch and metric scales to accurately measure and lay out linear distances

4. be able to use a protractor to accurately measure and lay out angles

5. know the types and meanings of the standard lines used in technical drawings

6. be able to apply the basic techniques of technical drawing to create clear and accurate instrument drawings of simple objects

2-2 DRAFTING MATERIALS

The following sections describe the materials most often used to draw on, draw with, and erase with.

Papers

Sketches may be made on any handy scrap of paper, but technical drawings are always done on high-quality papers called **drafting media**. Drafting media are resistant to tearing and abrasion and include **vellum** (paper made from cloth fibers), **tracing cloth** (linen), and **polyester film** (which is often referred to by the brand name **Mylar**). Each has its advantages, but the most commonly used media are vellum, because of its low cost, and polyester film, because of its toughness and dimensional stability. All these papers are semi-transparent. This property is useful for tracing and necessary for the production of copies by the diazo process. (Copying and blueprinting processes are described in Chapter 9.) Drafting media come in the standard sheet sizes shown in Table 2-1.

No matter what type of paper you are using, it should be fastened to your board with drafting tape while you are drawing with instruments. (Drafting tape is similar to masking tape, but less sticky. It is available in handy dispensers.) Align your paper's edges with the edges of your drawing board or table. Making sure that the sheet is lying flat, fasten down each corner securely with a small piece of tape as you will see in figures later in this chapter. Usually your paper should be positioned in the approximate center of your board.

Table 2-1 Standard Drafting Sheet Sizes

U.S. Sizes		Metric Sizes	
Designation	Size (in.)	Designation	Size (mm)
A	8½ × 11	A4	210 × 297
B	11 × 17	A3	297 × 420
C	17 × 22	A2	420 × 594
D	22 × 34	A1	594 × 841
E	34 × 44	A0	841 × 1189

Leads and Inks

Lines and figures are drawn with lead or ink. Drawing leads are made of graphite with clay or plastic resin added in varying amounts to make 18 different grades of hardness. They range from 9H (the hardest) to 7B (the softest).

Hard leads (grades 9H thru 4H): These are used for layout work, construction lines, guidelines, and where extreme accuracy is required.

Medium leads (grades 3H, 2H, H, F, and HB): These are used for general drawing, finished line work, and lettering. HB is about the same hardness as an ordinary No. 2 pencil.

Soft leads (grades B through 7B): These grades are too soft for most technical drawing, but they are widely used for artwork.

It is recommended that you use 6H lead for construction lines, 2H or H lead for finished line work, and H or F lead for lettering. With experience, however, you may decide that other grades suit your needs better. The suitability of the various grades of lead depends on individual drawing techniques.

The various grades of lead can be purchased as wood pencils or as lead sticks to be used in mechanical pencils (lead holders) and compasses. Mechanical pencils are available in two styles. Figure 2-1a shows the older

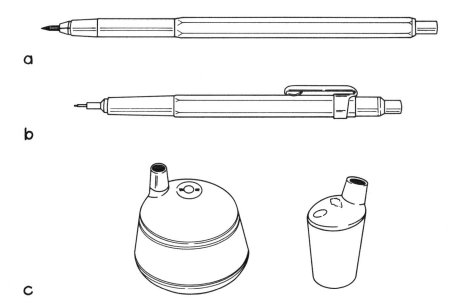

a

b

c

Figure 2-1 (a) An older type of mechanical pencil, or lead holder. (b) A fine-line mechanical pencil. (c) Two types of mechanical lead sharpeners.

Figure 2-2 (a) A correctly sharpened lead for a lead holder–type mechanical pencil. (b) and (c) Correctly sharpened leads for a compass.

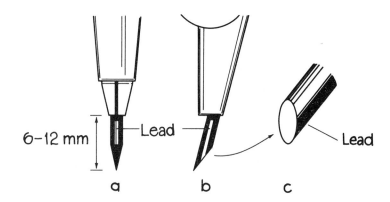

type, the lead holder, which uses the same large-diameter leads found in wood pencils. Like wood pencils, the leads in lead holders require frequent sharpening. Figure 2-1b shows the newer type of mechanical pencil, the fine-line pencil, or thin-line pencil. It holds small-diameter (0.3 mm to 0.9 mm) leads that never need sharpening and produce constant-width lines. Leads with diameters of 0.5 mm or 0.7 mm are the most commonly used. Fine-line pencils are rapidly replacing the old-style lead holders.

Wood pencils and large leads can be sharpened with a file or sandpaper, but mechanical sharpeners (shown in Figure 2-1c) are much more convenient and always produce a uniform point. Compasses (which are discussed in Section 2-3) use the large-size lead and should be sharpened to a flat chisellike point by using sandpaper. Always sharpen leads some distance from your drawing and other instruments. This prevents the graphite filings from coming in contact with them. Loose graphite is the primary cause of dirty drawings. Figure 2-2 shows correctly sharpened points for both large lead holders and compasses.

Technical drawing with ink is normally done with a technical fountain pen (shown in Figure 2-3) and fast-drying, waterproof black india ink. Technical pens are available with point sizes ranging from .13 mm to 2.0 mm in diameter. Inking can be done on any of the drafting media, but polyester film is preferred because mistakes made on it can be more easily corrected than mistakes on other media. Removal of ink from vellum and cloth is very difficult and seldom satisfactory. However, this drawback is more than com-

Figure 2-3 (a) A technical pen. (b) An adapter for use with a compass.

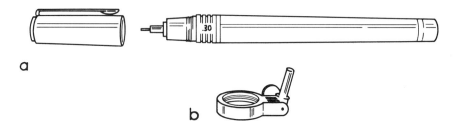

pensated for by the superior line quality possible with ink. Inking is done using the same instruments as used with lead; however, compasses require a special adapter (shown in Figure 2-3b) to accommodate technical pens. Pens also require special care and cleaning to prevent clogging or damaging of the precision points. Most manufacturers include directions for the care and cleaning of their pens, and the instructions should be carefully followed.

Erasers

Lead can be erased with any soft eraser. Commonly used types are the pink pearl and the soft white. A thin metal or plastic erasing shield (shown in Figure 2-4) may be used to protect line work adjacent to areas being erased. Use a dust brush, rather than your hand, to remove eraser crumbs and excess graphite from your drawing. Using your hand will smear the line work and

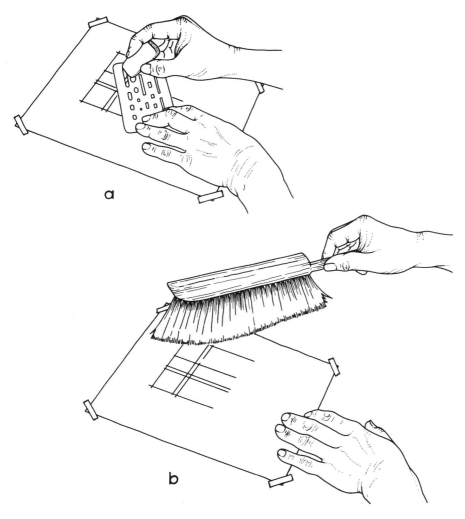

Figure 2-4 (a) Using an erasing shield. (b) Using a dust brush on a drawing sheet taped to the board.

lettering. Removing ink is sometimes a difficult task. Scraping it with a knife or razor blade is often the only method that works on vellum and cloth, but it is nearly certain that you will also scrape a hole in your paper. Ink can be removed from polyester film by using a vinyl eraser and ink-erasing fluid. (Saliva also works.)

2-3 DRAFTING EQUIPMENT

Many specialized tools and instruments are used in drafting. Because of the need for accuracy, most drafting tools are made of high-quality materials manufactured to very precise standards. Lower-quality tools are available, but they are seldom a bargain. It is difficult to achieve good results with them, and they seldom last very long.

Table 2-2 lists the equipment necessary for a beginning course in technical drawing. Many more tools are available, but those in Table 2-2 are sufficient for doing all the problems in this text. Figure 2-5 shows most of the basic tools.

Triangles, irregular curves, templates, and protractors are made from high-quality transparent plastic. The edges must be smooth and true. Scales are made from various materials, but the best scales are high-quality opaque plastic. The numerals and divisions should be engraved in the scale. Quality compasses and dividers are made from cast metal with chrome finishes and precision screws and joints. Drafting tables and drafting boards are generally made of soft wood with straight, smooth sides and top. It is best to cover the drawing surface with a vinyl cover.

Table 2-2   Basic Drafting Equipment*

45° triangle (6–8 in. sides)

30°-60° triangle (8–10 in. on the long side)

Scale (30 cm long with millimeter divisions and 12 in. long with .02-in. decimal divisions and ½₂-in. fractional divisions)

Protractor (3–4 in. 180° with ½° divisions)

Compass (6-in. legs)

Dividers (6-in. legs)

Circle template (with holes 2–30 mm or .06–1.25 in. in diameter)

Irregular (French) curve

Lettering guide

Dust brush

Erasing shield

Fine-line mechanical pencils (0.5 mm and 0.7 mm in diameter)

Drafting table or drafting board (20 in. × 24 in.)

*Cited sizes are suggestions.

Many other available drafting tools are not included in Table 2-2 because they are very expensive, have quite specialized purposes, or duplicate the functions of tools listed in the table. T-squares, parallel rules, and triangles

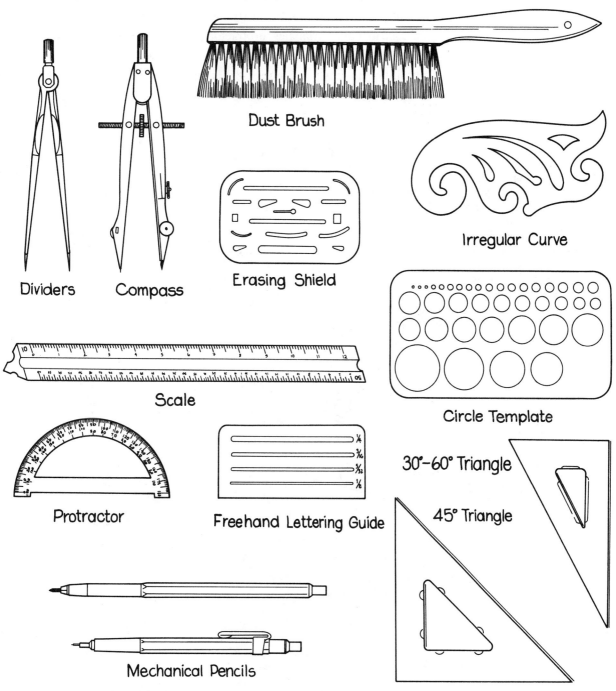

Figure 2-5 A drafting table or board and the tools shown here constitute the basic drafting equipment.

(shown in Figure 2-6) are instruments that perform the same functions as some of those in Table 2-2, but they are preferred by some drafters. T-squares and parallel rules are used to draw straight horizontal lines, and triangles are used to draw straight lines at various angles.

Most important of the expensive tools are **drafting machines**. Drafting machines are fairly complex instruments that are attached to drafting tables and used to replace triangles, protractors, T-squares, and parallel rules. There are two types of machines. The arm type is shown in Figure 2-7; the track, or rail, type is shown in Figure 2-8. On both types a head assembly holds two straightedge blades (or scales) rigidly mounted with a right angle between

Figure 2-6 The drafter slides the (a) T-square or (b) parallel rule up or down on the table or board to draw horizontal lines. Note that the parallel rule is attached to the drawing surface. Vertical lines are drawn with the triangles. Adjustable triangles are used to draw lines at various angles.

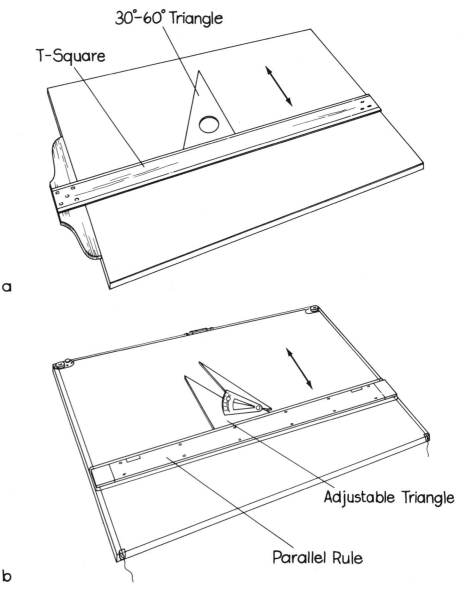

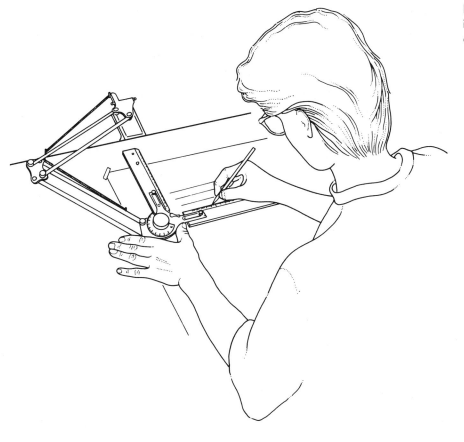

Figure 2-7 A drafter drawing horizontal lines with an arm-type drafting machine.

Figure 2-8 A drafter at work using a track-type drafting machine.

them. The entire head assembly can be rotated through 360° and moved, or translated, to any position on the drawing table. This makes drafting machines very versatile tools. They can be used to draw straight lines at any angle, parallel or perpendicular to any other line, and to measure lengths of lines and angles between lines. A well-built drafting machine, correctly used, can significantly increase the speed and accuracy of any drafter.

Of the specialized tools available, templates are the most important. **Templates** are pieces of thin plastic with various shapes cut out of them. They are used for drawing these shapes by tracing around the cutouts with a pencil or pen. Their use speeds up the drawing process considerably, and you should take advantage of them whenever possible. Some templates, such as circle templates (shown in Figure 2-5), have general utility, but most are designed for drawing various sizes of specific types of figures. There are, for example, templates for drawing ellipses, fasteners, electronic symbols, and architec-

Figure 2-9   Some of the specialized drafting templates available.

tural details. Examples are shown in Figure 2-9. For a full range of templates, consult suppliers' catalogs.

Using Straightedges

The drafting machine blades, the 45° triangle, and the 30°-60° triangle are called straightedges, for obvious reasons, and are used to draw straight lines. Horizontal, vertical, and inclined lines are drawn using a straightedge as a guide, as shown in Figures 2-7, 2-8, and 2-10.

When using a straightedge, place the point of your pencil against the edge of the guide, as shown in Figure 2-11a, holding it in the position shown in Figure 2-11b. (If you are left-handed, reverse the position of the hands and the 60° pencil angle.) To achieve a line of constant width and density, pull and rotate your pencil as you draw. Don't push your pencil—this causes the lead to dig in, which may result in tearing your paper. Rotating your pencil as you draw prevents the formation of a widened flat area on your lead and the subsequent widening of your line. If you are using an older type of mechanical pencil, it is especially important that you keep it sharpened. Failure to do so will result in lines that become very wide and fuzzy. This is not a problem with fine-line pencils and one reason why they are so widely used.

When drawing a line between two points, place your pencil on the paper at one of the points. Slide your straightedge so it touches the pencil lead. Using this as a pivot point, adjust your straightedge to the second point. Hold

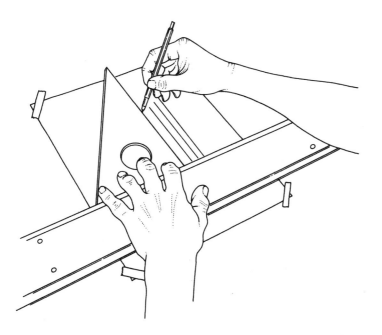

Figure 2-10 Drawing vertical lines by using a triangle.

Figure 2-11 (a) and
(b) The correct positioning of
a pencil used with a straightedge. (c) The correct position
of a technical pen used with
a straightedge.

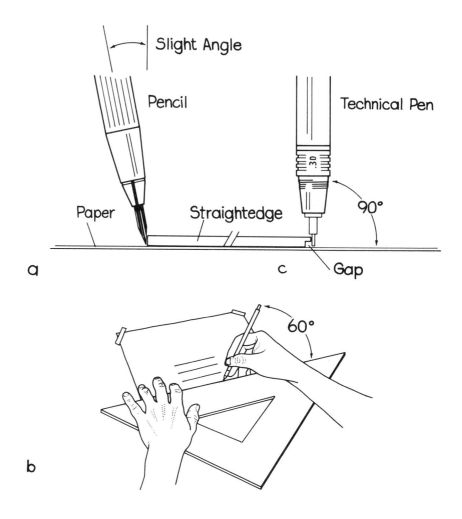

your straightedge firmly in place with one hand, and draw the line. When drawing light construction lines, one pass is enough, but for finished lines two or three passes are usually needed to ensure sufficient darkness.

Avoid sliding straightedges over completed lines—this will result in smeared line work. Plan your drawing so that, to draw the next line, you slide your instruments away from the last line drawn.

Drawing with a technical pen requires techniques different from those used when drawing with a pencil. Prior to drawing, gently shake your pen (away from your drawing) until you hear a clicking noise. This indicates that the cleaning needle is free and the ink passage is open. Try a few freehand lines on scrap paper to be sure. Set your pen point down directly at the beginning of each line, and start drawing immediately. Lift the pen straight up at the end of the line. Allowing the pen point to remain in contact with the paper at one point for any length of time can result in an ink blob. To ensure a steady flow of ink, the pen must be held nearly vertical while drawing. Keep your pen points wiped clean; otherwise, dried ink may build up and cause wide and irregular line work. Usually a single pass with your pen is sufficient to draw a dark solid line. Be sure to securely cap your pens when not in use.

When drawing with ink, straightedges and templates must be elevated a millimeter or so off the paper or the ink will run under the instruments and ruin your drawing. You can elevate the instruments by laying one on top of another or by applying several layers of drafting tape a short distance from their edges. As Figure 2-11c shows, instruments with recessed edges are available for use when inking. The recessed edges serve the same purpose as elevating the instruments.

As shown in Figure 2-12, triangles may be used singly or in combination to draw straight lines at angles of varying degrees in regard to a given line. Triangles are also used to draw lines parallel or perpendicular to a given line.

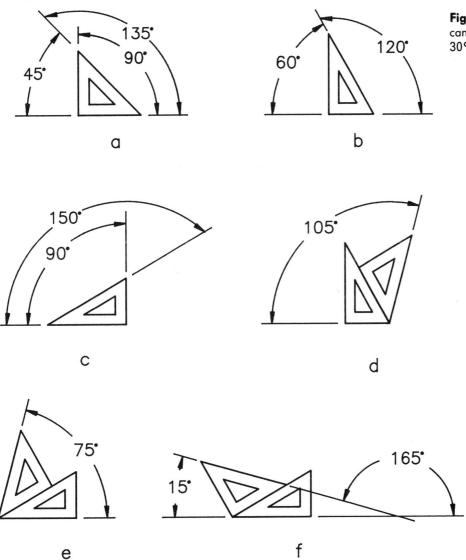

Figure 2-12 Angles that can be drawn using 45° and 30°–60° triangles.

Figure 2-13 Parallel and perpendicular lines may be drawn by using two triangles. Holding the lower triangle steady, slide the upper one from the original to the second position. Lines drawn along the edges of the upper triangle will be either parallel or perpendicular to lines drawn at the original position.

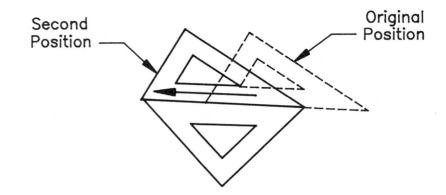

Begin by aligning one side of a triangle with a given line. Place the other triangle so that it lies against the first one, as shown in Figure 2-13. Holding the second triangle firmly in place with one hand, move the first triangle to the desired position by sliding it along the edge of the second triangle. You may then draw lines perpendicular or parallel to the original line.

Using Compasses and Circle Templates

Either the compass or the circle template may be used to draw circles and arcs. The compass lead (use an H or F grade) should be positioned and sharpened as shown in Figures 2-2 and 2-14. Make sure the lead and the shoulder point extend the same distance from each leg. To use a compass, set the distance between the points equal to the radius (R) of the circle, or arc, by measuring against a construction line drawn to the correct length or to a scale (see Figure 2-15). Always set the distance by opening the compass out farther than required and then closing it back to the correct distance. This procedure eliminates any looseness, or play, in the screw and provides a firm setting. Then

Figure 2-14 (a) The proper position for a compass lead. (b) The proper sharpening of a compass lead.

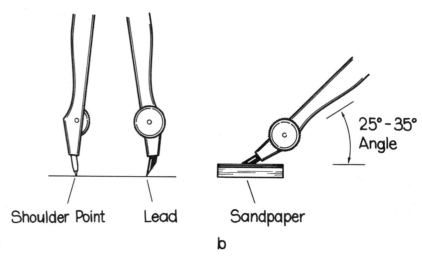

Shoulder Point   Lead   Sandpaper

a   b

place the compass point directly on the previously located circle center, and firmly push the point into the paper. Lower the lead to the paper and rotate it completely around the center at least once in each direction, holding the compass with a slight lean in the direction of movement (see Figure 2-16).

Circle templates are flat, thin pieces of plastic with holes of various sizes cut into them. The holes are labeled by their diameters ($\emptyset$, or D). The standard metric template has 30 to 40 holes ranging in diameter from 2 mm to 30 mm. The standard template with U.S. units has 30 to 40 holes ranging in diameter from .062 in. to 1.25 in. Templates with larger holes are available. To use the template, select the hole of the proper size and line up the index marks printed on the edges of the hole with the centerlines previously drawn on the paper. Use a sharp pencil and trace around the hole cutout as shown in Figure 2-17. Whenever possible, use the circle template rather than a

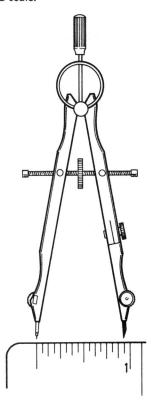

Figure 2-15 Setting a radius by measuring against a scale.

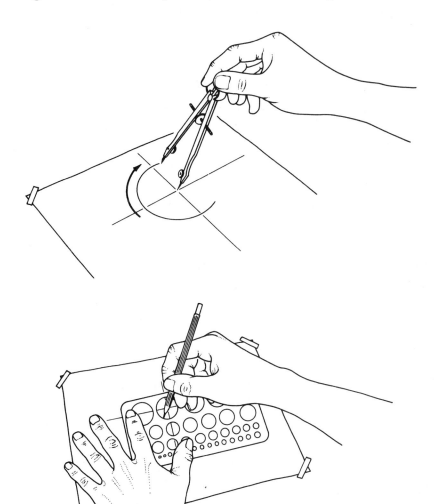

Figure 2-16 Drawing a circle or arc with a compass.

Figure 2-17 Drawing a circle or arc with a circle template.

Figure 2-18 Using an
irregular (French) curve.

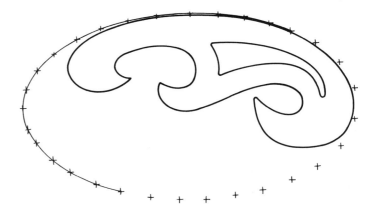

compass; it can be less accurate, but it's much easier and faster to use. When drawing arcs dimensioned by their radii, remember to double the radius to find the correct hole on the template.

Using Irregular Curves

Curved lines that cannot be drawn with a compass or circle template are drawn by using the irregular (French) curve. You must first establish at least three points through which the curved line will pass. Adjust the irregular curve by trial and error to accommodate as many consecutive points as possible. Draw a line through these points. Move the instrument to a new position that accommodates additional points, and repeat until the entire curve is completed. It is always important to overlap the previously drawn line when setting up the instrument in its new position, as shown in Figure 2-18. This will ensure a smooth continuous curved line with no bumps, or obvious starting and stopping points.

2-4 MEASURING EQUIPMENT

To make drawings that are the correct size, you must use measuring equipment. The principal measuring devices are scales, for making linear (straight-line) measurements, and protractors, for making angular measurements.

Linear Measurements

Drafting scales are the instruments used for making linear measurements. These instruments are made with either flat or triangular cross sections and come in 15-cm, 30-cm, 6-in., and 12-in. lengths. Each instrument has two or more different scales imprinted on it. Figure 2-19 shows examples of some commonly used drafting scales.

To read a scale, you must understand the meaning of the term *least count*. **Least count** is the value, or distance, between any two closest marks on any scale. In other words, least count is the smallest distance you can measure with certainty, using a given scale. However, skilled drafters can make consistently good estimates of one-half the least count.

You must also understand the term *units*. **Units** are the basic divisions imprinted on each scale, such as millimeters or inches. For example, the standard mechanical engineer's scale used in the United States has inches for its units. Each inch is divided into 50 equal parts, making the least count 1/50, or .02, of an inch. The standard metric scale is divided into centimeters, and each centimeter is divided into 10 mm. Thus, the least count and the units are both millimeters on the full-size metric scale.

It is preferable, but not always possible, to draw objects full-size, that is, the same size they actually are. Objects that are too large to fit on a sheet of paper when drawn full-size (such as a building or an airplane) must be drawn at a reduced size. Objects that are very small (such as a watch gear or a

a. Metric Scale

b. Engineer's Scale

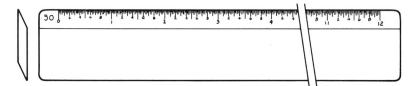

c. Decimal=inch Scale

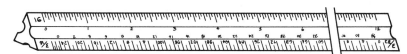

d. Architect's Scale

Figure 2-19 Examples of some commonly used scales.

Table 2-3 Typical Reduction and Enlargement Scales

Reduction	Meaning
1/2; 1:2, half	The drawing is one-half the size of the object. 1 mm or 1 in. on the drawing equals 2 mm or 2 in. on the object.
1/4, 1:4, quarter	The drawing is one-quarter the size of the object.
1/10; 1:10	The drawing is one-tenth the size of the object.
1 in. = 50 ft	One in. on the drawing equals 50 ft on the object.
1 in. = 100 ft	One in. on the drawing equals 100 ft on the object.
1/4 in. = 1 ft 0 in.	One-quarter in. on the drawing equals 1 ft on the object.

Enlargement	Meaning
2×, 2:1, twice	The drawing is two times the size of the object. 2 mm or 2 in. on the drawing equals 1 mm or 1 in. on the object.
10×, 10:1	The drawing is ten times the size of the object.

microchip) must be drawn at an enlarged size. Therefore, drawings are often smaller or larger than the objects they depict.

The ratio of the drawing size to the object size is called the **scale** of the drawing. Reductions and enlargements are always proportional. For example, on a half-size drawing, the object is drawn one-half its actual size. Table 2-3 lists some typical reductions and enlargements. To facilitate the drafting of reduced and enlarged drawings, a number of scales are available that can be used to measure directly in the desired units.

Metric scales.   **Metric scales** are designed to be used for both full-size and reduced-size drawings. The basic unit of length in the metric system is the meter (approximately equal to 39.37 in.). The full-size metric scale (shown in Figure 2-20a) is divided into centimeters (30.5 cm = 1 ft) and millimeters (1 cm = 10 mm) and is labeled "cm," "mm," or "1:100." The least count of this scale is 1 mm, and the units are millimeters. This scale, then, can be used for full-size drawings and also for reduced-size drawings such as 1/10, 1/100, and so on. There are many other reduced-size metric scales, the most common having ratios of 1:80, 1:50, 1:40, 1:33.3, and 1:20. The corresponding scale factors are shown with each scale to indicate what decimal part of a meter each numbered division represents. For example, on the 1:40 scale in Figure 2-20b, the ".025" indicates that each numbered division is .025 meters long.

Full-size inch scales.   **Full-size inch scales** are divided either fractionally (1/32, 1/16, and so on) or decimally (.02, .10, and so on) and are generally used

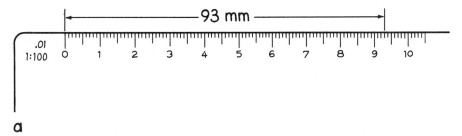

a

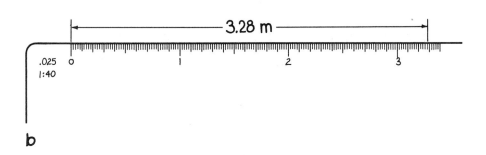

b

Figure 2-20 (a) A metric scale for making full-size (for example, 93 mm) or 1/100 reduced-size (for example, 9.3 m) drawings. (b) A metric scale for making 1/40 reduced size (for example, 3.28 m) drawings.

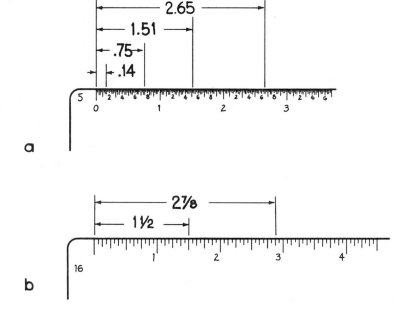

a

b

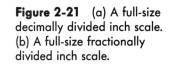

Figure 2-21 (a) A full-size decimally divided inch scale. (b) A full-size fractionally divided inch scale.

for making full-size drawings of manufactured products. The least count of the decimal scale shown in Figure 2-21a is .02 in., indicated by the label "50." The "50" means that each inch is divided into 50 equal parts. The least count of the fractional scale shown in Figure 2-21b is 1/16 in., indicated by the label "16." The units for both scales are inches. Half- and quarter-size inch scales are also available.

Figure 2-22 (a) An engineer's scale, with 10 divisions per inch, being used for drawing 1 in. = 100 ft reduced-size drawings. For example, 2.60 in. on the drawing represent 260 ft on the object. (b) An engineer's scale, with 20 divisions per inch, being used for drawing 1 in. = 20 m reduced-size drawings. For example, 1.10 in. on the drawing represent 22 m on the object.

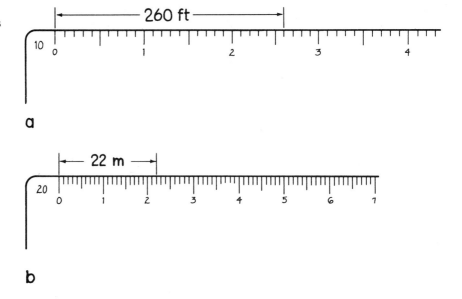

a

b

Engineer's scales. **Engineer's scales** are divided such that each inch has an equal number of parts (10, 20, 30, 40, 50, and 60). Each of these six scales is used for making reduced-size drawings of large structures and for plot plans, maps, and other civil engineering projects. The drawings are made to scales such as 1 in. = 10 ft, 1 in. = 300 yd., 1 in. = 50 mi, and so on—scales where each inch on the drawing represents some longer distance on the real thing. (In practice the scale equivalents are often expressed in forms using the symbols for inch and feet—1″ = 10′, for example.) The units and least count may be whatever the user needs. The 10 scale, as used in Figure 2-22a, is measuring in feet and has a least count of 10 ft; the 20 scale shown in Figure 2-22b is measuring in units of meters and has a least count of 1 m.

Architect's scales. **Architect's scales** have one face divided into inches and subdivided into 1/16 in. The other faces are designed for making reduced-size drawings from ¼ size (3 in. = 1 ft 0 in.) to $\frac{1}{128}$ size (3/32 in. = 1 ft 0 in.). The faces are divided so that the foot, as the basic unit, is represented by some fractional value of a foot. The standard foot distance has been compressed into a shorter distance by proportionally shorter inches and fractions of inches. These scales are primarily used for making architectural and structural drawings. Figure 2-23a shows 3/4 in. = 1 ft 0 in.; Figure 2-23b shows 1/8 in. = 1 ft 0 in. The architect's scale is called an **open divided scale** because only one portion on the end of each scale is subdivided into the scale's least count.

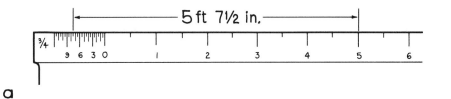

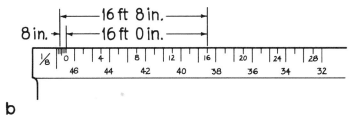

a

b

Figure 2-23 (a) An architect's scale for making 3/4 in. = 1 ft 0 in. reduced-size drawings. (b) An architect's scale for making 1/8 in. = 1 ft 0 in. reduced-size drawings.

Scaling Techniques

You can use scales either to measure distances between two existing points or to lay out known distances on your drawing. In either case follow these rules:

1. Always start by adjusting your scale so that the zero mark is centered on the first point, or mark. Note: The architect's scale is an open divided scale and read in a different manner than the others. Start by locating the nearest whole foot mark in your measurement and aligning it with the first point. Read the second point in the fully divided portion at the beginning of each scale. See Figure 2-23 for examples.

2. Always measure to the center of lines.

3. Always measure along a line. If there isn't one, then draw a line in the direction you wish to measure and align your scale with it.

4. Never use your scale as a straightedge for drawing lines.

When you have a space to be divided into equal or proportional parts, a technique known as **slant scaling** can save you much time and help eliminate mistakes. In Figure 2-24, for example, two lines are drawn 93.7 mm apart. To divide this space into two equal parts by normal methods requires dividing 93.7 by 2, drawing a line perpendicular to the two given lines, and then measuring 46.85 mm (which is not easy to do) to the middle. Using the slant scale technique, select a distance greater than 93.7 that is easily divided by 2. In this case 100 mm, half of which is 50 mm, would be appropriate. Adjust the

Figure 2-24 The slant scaling technique for dividing a distance into equal parts.

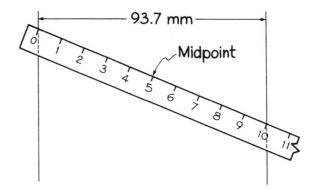

scale so that the "0" is on one line and "10" on the other. Obviously 50 mm, or "5" on the scale, is in the middle. You can use this method to divide any space into any number of parts by selecting a distance easily divisible by that number.

As shown in Figure 2-25, dividers are used to set off or transfer distances already established on a drawing. This is a more accurate process than transferring distances with your scale, simply because you do not have to read the dividers, thus eliminating an opportunity for making a mistake. Set each divider point on opposite ends of the distance to be transferred. Move your instrument to the new location and push the points gently into the paper. Mark the end points with your pencil. Take special care not to change your divider's setting while making the transfer. If you do not have any dividers, you can transfer distances with your compass, but not quite as accurately.

Figure 2-25 Using dividers to set off equal distances along a line.

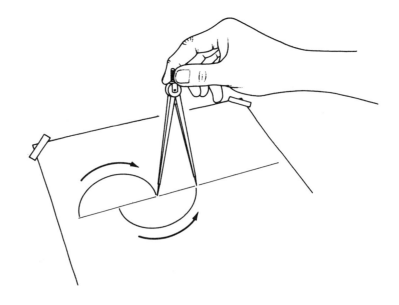

It is often necessary to convert measurements from millimeters to decimals of an inch and to fractions of an inch. To facilitate these conversions, conversion charts such as Table A-1 and Table A-2 in the appendix are used. Two very useful conversion factors to remember are 1 in. = 25.4 mm and 1 mm = .04 in. (approximately).

Angular Measurements

Protractors (shown in Figure 2-26) are used to measure angles. Angles, in most technical drawings, are measured in degrees. As you probably know, a complete circle contains 360°. So, all angles in technical drawings are between 0° and 360° degrees. Note that the size of an angle doesn't change when the scale of a drawing changes.

The edges of protractors are commonly divided into degrees and half degrees (30 minutes). To measure an angle, place the center mark of your protractor on the origin of the angle and align the zero mark with one leg of the angle. Read the size of the angle where the other leg intersects your protractor edge. See Figure 2-27 for examples.

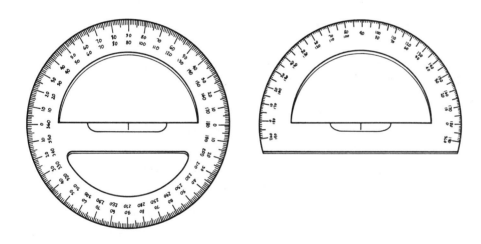

Figure 2-26 Two different styles and sizes of protractors.

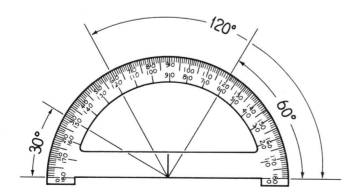

Figure 2-27 Angular measurements are made with a protractor.

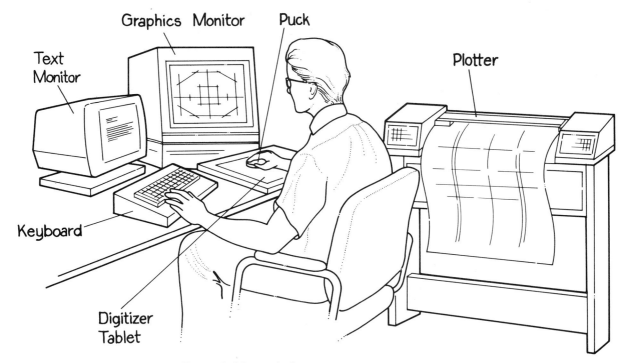

Text
Monitor

Graphics Monitor

Puck

Plotter

Keyboard

Digitizer
Tablet

Figure 2-28 A drafter operating a computer-aided drafting (CAD) system.

2-5 COMPUTER-AIDED DRAFTING SYSTEMS

Traditional drafting tools are being replaced in industry by **computer-aided drafting (CAD)** systems. CAD systems consist of computers connected to special input devices such as digitizers and output devices called plotters. The basic CAD components are shown in Figure 2-28. CAD systems are used by drafters to create very accurate and uniform drawings in much shorter times than with traditional tools. Drawings can be stored in computer memory, displayed on the computer screens, or output on paper by means of plotters. CAD systems are powerful and efficient when operated by competent drafters who thoroughly understand the principles of technical drawing.

2-6 LINE TYPES

Many types of lines are used in technical drawing. Each one has a specified form and a specific and unique meaning. These forms and meanings are published and maintained by the American National Standards Institute (ANSI).

The most commonly used standard lines are described and illustrated in Figure 2-29. Figure 2-30 shows some typical applications of these lines.

1. Visible object line (thick)

Solid unbroken lines are used to show edges and outlines of objects.

2. Hidden object line (thin)

Dashed lines are used to show hidden edges and hidden outlines of objects. Dashes should be 3–6 mm or .10–.25 in., depending on the scale used. In most drawings, gaps between dashes should be 1 mm or .03 in.

3. Centerline (thin)

Dashed lines are used to show symmetry or location of objects. Dashes should be of alternating lengths: 12–24 mm or .50–1.00 in. for long dashes and 3–6 mm or .10–.25 in. for short dashes. Gaps between dashes should be 1 mm or .03 in.

4. Phantom line (thin)

Dashed lines are used to show alternate positions and the outlines of reference objects. Dash sizes and spacing for phantom lines are the same as those for centerlines, except two short dashes are used instead of one.

5. Dimension, extension, and leader lines (thin)

Solid lines are used to show dimensions and their extents. Dimension lines and leader lines usually have arrows at their ends.

6. Cutting or viewing plane line (thick)

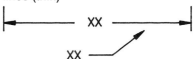

Extra-thick dashed lines with large arrows are used to indicate the location of a cutting or viewing plane for section or auxiliary views. Lines may consist of one long and two short dashes or dashes of equal length.

7. Break lines (Short are thick; long are thin)

A freehand line is used to show short artificial breaks in objects. A solid line with zigzags at 25-mm or 1-in. intervals is used to portray long breaks.

Figure 2-29 Standard line types. Thick lines should be 0.7 mm or 0.032 in., depending on the scale used. Thin lines should be 0.35 mm or 0.016 in.

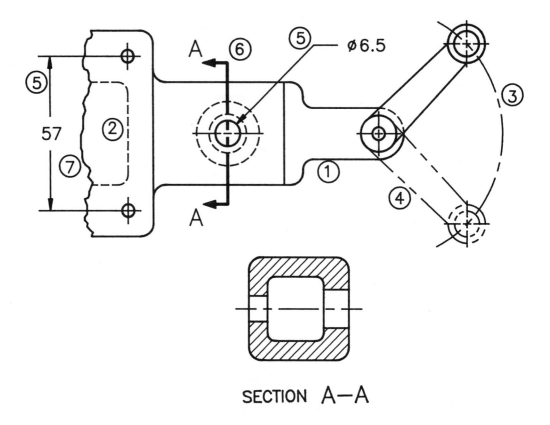

SECTION A—A

Figure 2-30 Examples of standard line types in use. Circled numbers refer to the line types shown in Figure 2-29.

2-7 DRAWING TECHNIQUES

The drafting of technical drawings has three principal objectives. In order of importance these are (1) clearness, (2) accuracy, and (3) speed. The first two come about through the use of proper technique, and the last through experience and practice. Some basic techniques that will help you achieve these three objectives follow.

1. Plan ahead. Before you start your instrument drawing, do a quick freehand sketch of the problem to determine which view(s) to draw.

2. Determine the scale and select the proper type and size of paper.

3. Position your paper in the approximate center of the drawing board, and tape it down, with the paper's edges parallel to the board's edges.

4. Locate the center of the paper with light diagonal lines.

5. If the object is symmetrical, draw its centerlines and set off vertical and horizontal lengths.

6. Block in the basic outlines of the object with construction lines. Use a 6H or harder lead and a very light pressure on your pencil.

7. Draw in the detail with construction lines.

8. Examine your drawing for errors and omissions.

9. If you are satisfied that your drawing is correct and complete, draw and darken the arcs and circles with a medium-grade lead (2H or H) or with ink.

10. Darken all straight lines. Lines should be dark, shiny, and sharp — not faint and fuzzy. This may require going over the lines two or three times with your pencil.

11. As shown in Figure 2-31a, make all corners sharp, with no overlaps or gaps. This is best achieved by starting at the corner and drawing away from it.

12. As shown in Figure 2-31b, make all transitions between arcs and straight lines smooth and even. This is best achieved by drawing the arcs first and the straight lines last.

13. Draw centerlines as required.

14. If necessary, erase any construction lines.

15. Do any lettering and draw borders and the like after the picture is completed.

16. Keep drawings clean by:

   • frequently washing instruments with mild soap and water

   • keeping hands clean

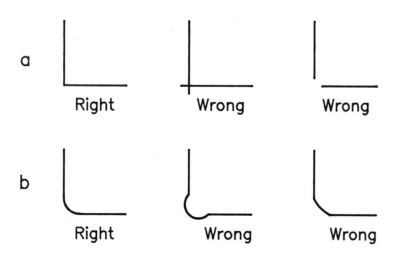

Figure 2-31 Right and wrong ways to draw corners and tangencies.

- not sliding instruments over previously drawn lines

- using a brush rather than your hand to clean erasings off drawings

- not sharpening leads, eating, or drinking near drawings

- covering completed portions of drawings with scrap paper while lettering or working on other portions

Figure 2-32 illustrates the application of some of the preceding steps in constructing a technical drawing.

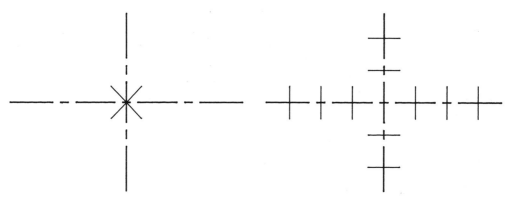

Step 1—Locate &
draw centerlines.

Step 2—Set off horizontal
& vertical lengths.

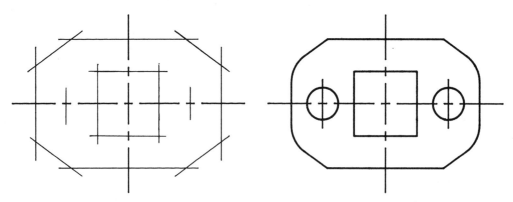

Step 3— Draw straight
line segments with
construction lines.

Step 4— Draw circles &
arcs, darken straight lines,
& erase construction lines
as needed.

Figure 2-32   Some of the steps to be followed in constructing a technical drawing.

2-8 CONCLUSION

This chapter has covered the basic materials and equipment used in making technical drawings. The most commonly used papers are vellum and polyester film. Lines may be drawn in pencil or ink. Triangles and drafting machines are used to draw straight lines; circle templates, compasses, and irregular curves are used to draw curved lines. Scales are used to make straight-line measurements, and protractors are used to measure angles. All these materials and tools should be of high quality to produce high-quality drawings. Cheap tools and supplies are difficult to work with. Of course, the quality of your materials and equipment will never be any better than your ability to use them. For that reason the drawing techniques presented in this chapter need to be followed and practiced.

This chapter also presented the standard lines and their meanings. It is important to know their forms and meanings because you will be using them throughout the remainder of this book. In a way, the standard lines are the basic vocabulary of technical drawing. You must use them correctly to communicate through your drawings.

Check your understanding of this chapter by answering the following questions.

REVIEW QUESTIONS

1. To draw, with construction lines, an object that measures 230 mm × 340 mm you would need:

   a. 6H lead and A4 paper

   b. 2H lead and A4 paper

   c. 6H lead and A3 paper

   d. F lead and A3 paper

2. The most dimensionally stable drafting media is:

   a. vellum

   b. polyester film

   c. tracing cloth

   d. onion skin

3. An easy way to draw parallel and perpendicular lines is by using:

   a. a triangle and a scale

   b. a scale and a compass

   c. two scales

   d. two triangles

4. Curved lines are drawn using:

a. circle templates

b. compasses

c. irregular curves

d. all the above

5. The least count of a scale is:

a. the smallest division on the scale

b. the number of divisions on the scale

c. the same as the units for any scale

d. the smallest unit on the scale

6. Which of the following is *not* a common type of scale?

a. metric

b. isometric

c. engineer's

d. architect's

7. The decimal-inch scale has _____ for units and a least count of _____.

a. inches; .02

b. inches; .01

c. feet; .02

d. feet; .01

8. The inch equivalent for 20 mm is approximately _____, and the millimeter equivalent for .375 in. is approximately _____.

a. 0.81; 9.5

b. 0.79; 9.5

c. 0.77; 9.1

d. 0.79; 8.7

9. Visible and hidden are both types of _____.

a. section lines

b. phantom lines

c. object lines

d. break lines

10. Lines used to indicate symmetry are called:

a. phantom lines

b. center lines

c. section lines

d. dimension lines

11. An important objective of drafting is:

    a. speed

    b. clearness

    c. accuracy

    d. all the above

12. It is best to lay out a drawing with construction lines before darkening any line because:

    a. it takes less graphite

    b. mistakes are easy to erase

    c. it's easier to keep the pencil sharp

    d. none of the above

PROBLEMS

2-1 On an A-sized sheet draw a border and title block as shown in the illustration. Fill in the title block by lettering the information indicated. Use uppercase letters 3-mm high or .12-in. high. Use guidelines and center vertically in the spaces. Find the center of the drawing space inside the borders by drawing diagonal lines between corners. Dimensions are in inches.

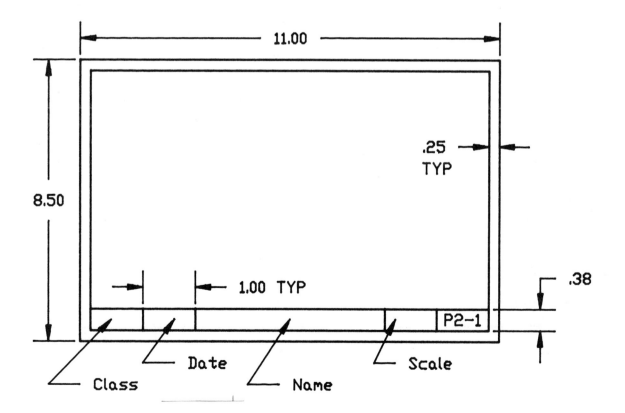

2-2   Measure each line by using the scale indicated. Write the measurements on a separate sheet of paper.

Use the cm scale; measure in millimeters.

a _____

b _____

c _____

d _____

e _____

f _____

g _____

h _____

Use the decimal inch (50) scale; measure to .01 in.

l _____

j _____

k _____

l _____

m _____

n _____

o _____

p _____

Use the fractional inch scale; measure to 1/32 in.

q _____

r _____

s _____

t _____

2-3 On a separate sheet of paper, draw lines as shown. Measure the indicated lengths, full-size; then measure the same lengths at half size. Mark and identify each distance.

35 mm |———————————————————————————————

152 mm |———————————————————————————————

97 mm |———————————————————————————————

7.5 mm |———————————————————————————————

115 mm |———————————————————————————————

139 mm |———————————————————————————————

62.5 mm |———————————————————————————————

41 mm |———————————————————————————————

2.5 mm |———————————————————————————————

1.55 mm |———————————————————————————————

5.10 In. |———————————————————————————————

.51 In. |———————————————————————————————

3.22 In. |———————————————————————————————

2.68 In. |———————————————————————————————

4.68 In. |———————————————————————————————

.14 In. |———————————————————————————————

3 3/4 In. |———————————————————————————————

1 1/8 In. |———————————————————————————————

5 1/16 In. |———————————————————————————————

4 7/32 In. |———————————————————————————————

draw blue guide line

2-4 Measure each line by using the scale indicated. Write the measurements on a separate sheet of paper.

write measure

-1In.=10ft ————————————————

-1In.=100ft ———————————————

1In.=20ft ——————————————————————————

1In.=300ft —————————

1In.=40ft ————————————————————————————

1In.=50ft ————————————————————————

1In.=6000ft ——————————————————————————————

-1cm=10m ——————————————————————

-1cm=100m ———————————————

1mm=1m ————————————————————

1/8'=1'-0 ————————————————————————————————

-1/4'=1'-0 ————————————————————

3/8'=1'-0 ——————————————————————————————————

-1/2'=1'-0 ——————————————————————————

3/4'=1'-0 —————————

1'=1'-0 ————————————————————————————————

3/32'=1'-0 —————————————————————

3/16'=1'-0 ——————————————————————————————

1 1/2'=1'-0 ————————————————————————————————————

3'=1'-0 ——————————————

2-5 Measure the angles indicated to the nearest degree. Write the measurements on a separate sheet of paper.

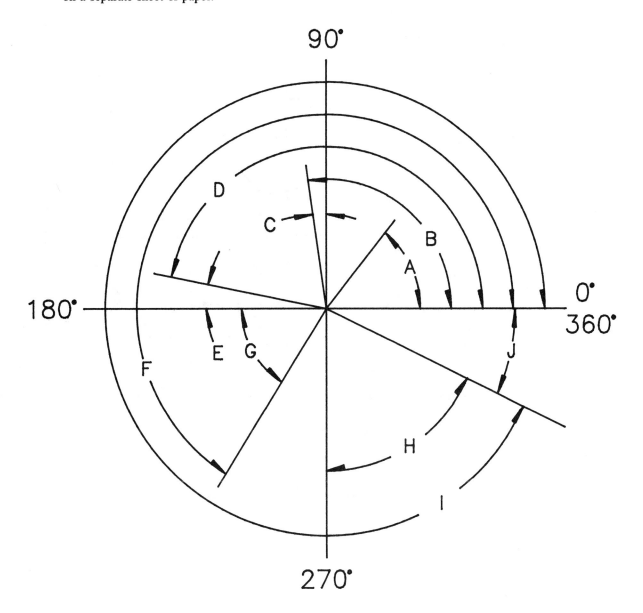

2-6 Measure the indicated linear dimensions in both millimeters and inches and the angular dimensions in degrees. Enter the measurements in a table on a separate sheet of paper.

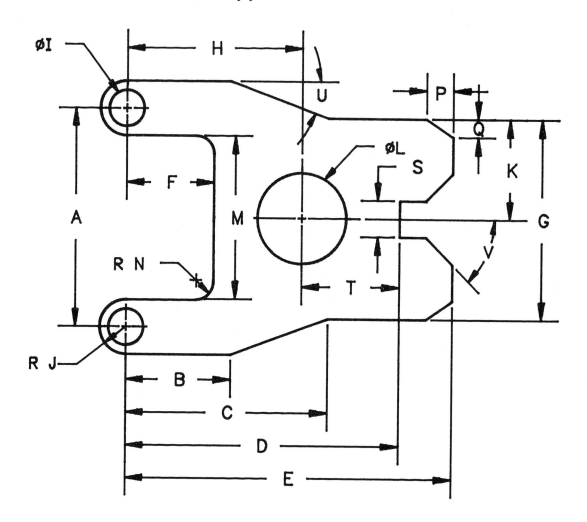

2-7 Redraw figures a through d on graph paper by using the scale 1=1 square.

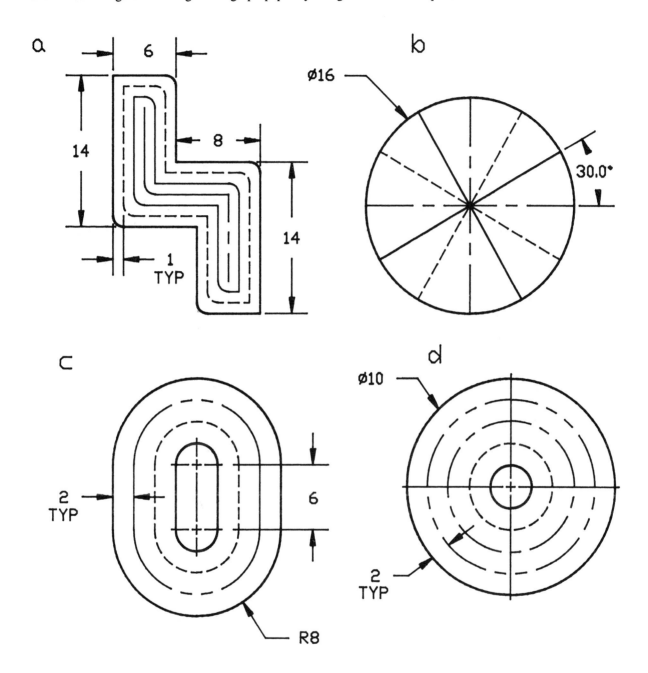

2-8   Redraw figures a through d as assigned. Dimensions are in inches.

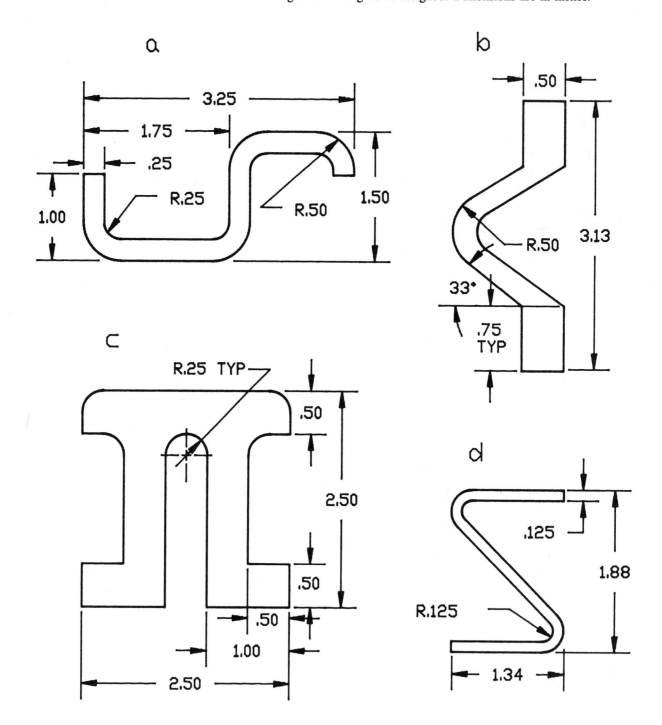

2-9 Redraw figures a through d. Dimensions are in millimeters.

a

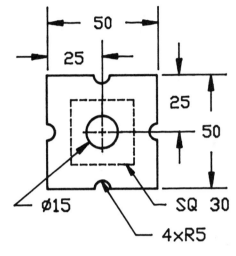

b

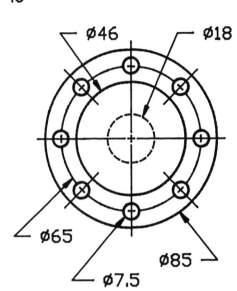

c

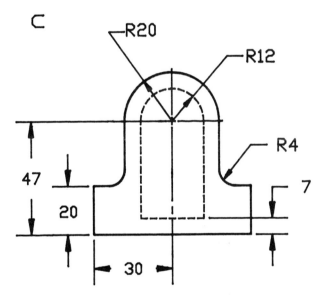

d

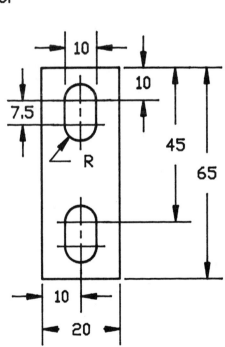

2-10 Trace the given points on a sheet of vellum and connect them with smooth curved lines.

a

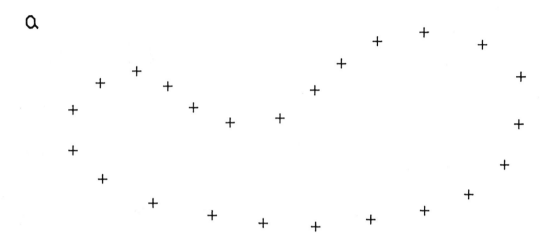

b

2-11   Redraw figures a and b by using the lengths and angles given.

a. Find the true size of angle a. All lines are the same length.

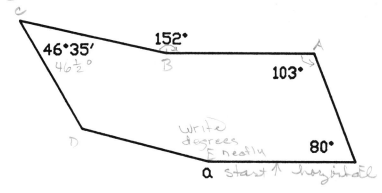

position work in
center

construction lines 1st
until all lines drawn
then darken

46°35'
46½°

152°

103°

80°

Write
degrees
neatly

a   start horizontal

b. Find the true size of angle a and the true length of line ab.

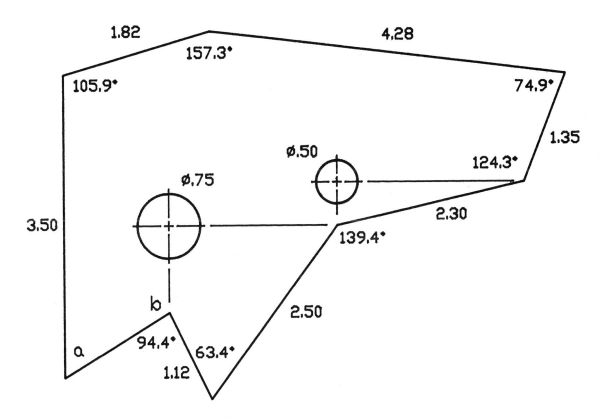

2-12 Redraw these pictorials full size on A-sized paper. Scale the figures for dimensions.

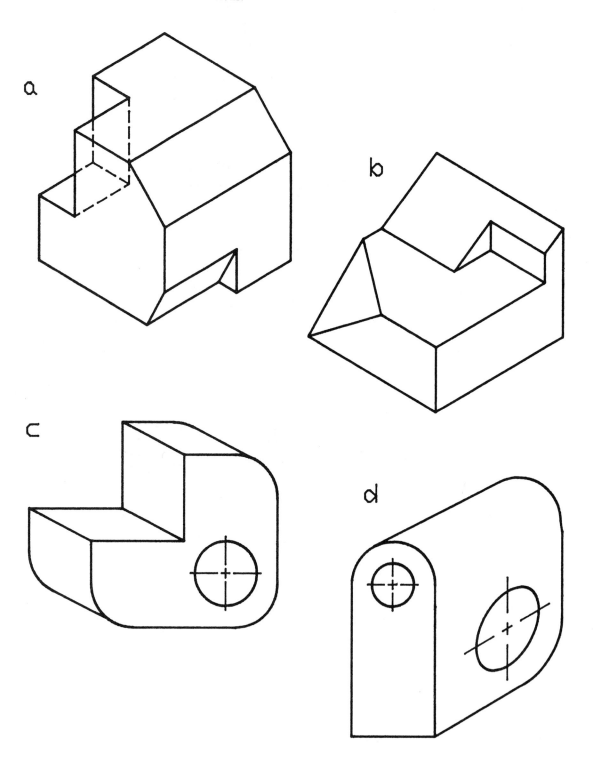

2-13 Redraw the multiview full size on A-sized paper. Dimensions are in inches.

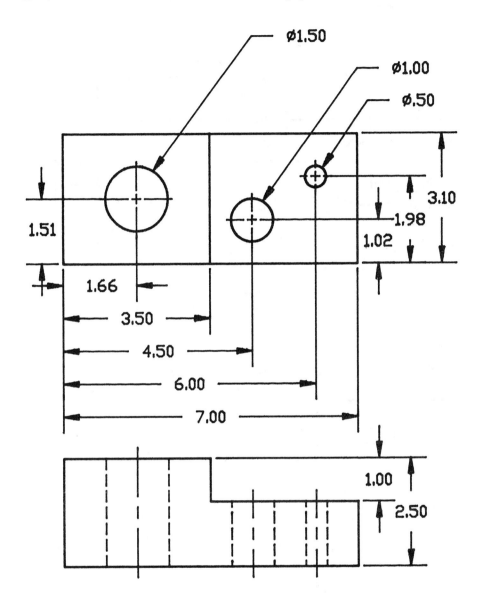

2-14 Redraw the multiview full size on A-sized paper. Dimensions are in millimeters.

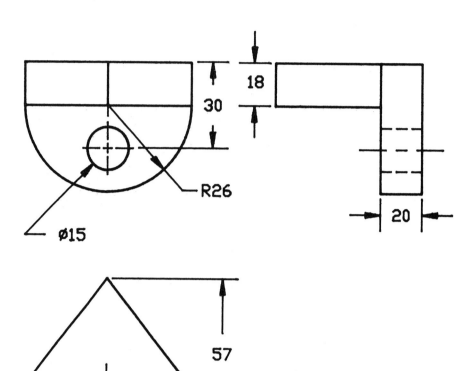

2-15   Redraw the multiview on A-sized paper. Scale the drawing for dimensions.

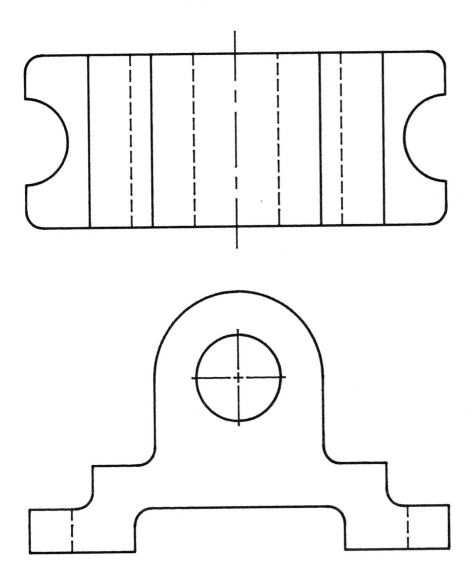

2-16 Redraw these figures on A-sized paper. Scale the figures for dimensions.

2-17   Redraw these figures on any sized paper.

a

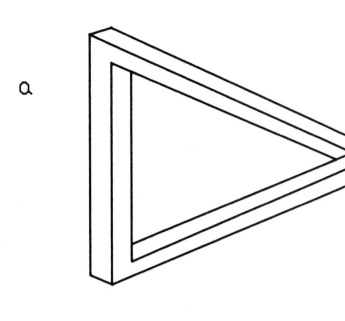

b

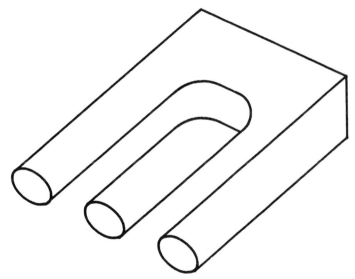

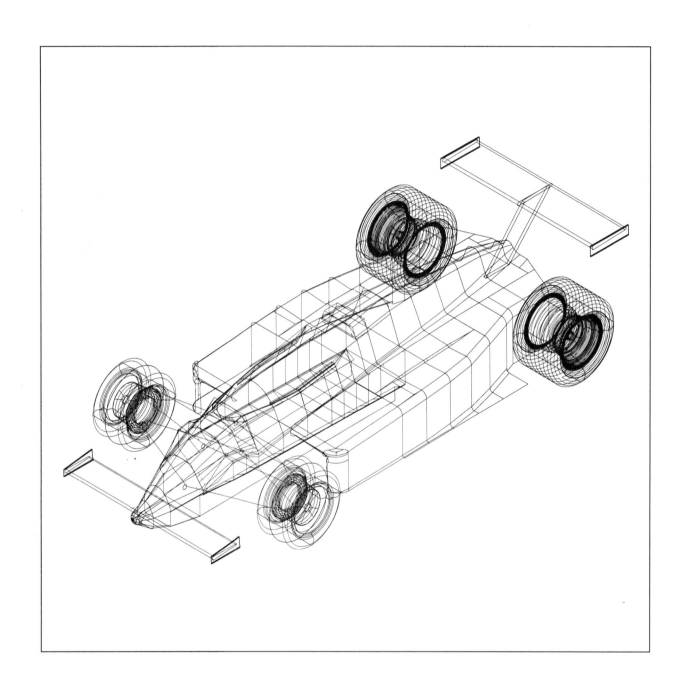

GRAPHICAL GEOMETRY
AND CONSTRUCTIONS

3-1 INTRODUCTION

Technical drawing makes extensive use of graphical geometry and geometric construction techniques. The illustration that opens this chapter shows a racing-car design drawn with a computer-aided drafting system. The design is made up of many geometric shapes: straight lines and curved lines, squares, triangles, circles, cubes, cylinders, and cones. Have you ever wondered how designers create things like new cars? As you probably imagine, it is a very complex process. One of the most useful and important tools designers have is technical drawing based on graphical geometric principles. Designers solve many of their problems by using geometric construction techniques.

A thorough understanding of graphical geometry and construction is necessary for a basic understanding of technical drawing. It is essential to know the geometric terms and relationships widely used in technical drawing, the techniques for placing points and lines, and the techniques for constructing geometric figures. These concepts will be used throughout the remainder of this book.

This chapter will present basic geometric construction techniques, provide fundamental definitions of geometric terms, and give you an opportunity to use your drawing instruments. The instruments required to do this work are two triangles, a compass, a scale, and a sharp hard (4H to 6H) pencil. In the examples in this chapter, constructions are indicated by thin lines, the given (starting) information and solutions by thick lines, and other information by dashed lines.

After you complete this chapter, you should:

1. know the geometric terms and definitions relating to straight lines and circles

2. be able to bisect straight lines, curved lines, and angles

3. be able to calculate the circumferences and areas of circles

4. be able to construct and locate tangents between curved and straight lines and between two curved lines

5. be able to calculate the areas and angles of plane figures

6. be able to construct a variety of common plane figures, including regular polygons

7. be able to construct approximate ellipses

8. know the basic solid geometric forms

3-2 THE GEOMETRY OF STRAIGHT LINES

Definitions

A **straight line** is any line that continues in the same direction for its entire distance. Straight lines, in technical drawing, are defined by their length and direction. Length is measured in inches, millimeters, feet, meters, miles, or the like. Direction is identified by referencing the line to some other line or a plane. For example, **vertical lines** have an up-and-down direction. Though it is not exactly true, it is convenient and useful to think of a vertical line as being **perpendicular** (90°) to the earth's surface. A **horizontal line,** also called a **level line,** is perpendicular to the vertical line. Therefore, the earth's surface is considered horizontal, or level, and any line lying on it is also horizontal. Any straight line that is not vertical or horizontal is called **inclined.** You can apply these ideas by thinking about the house shown in Figure 3-1. The sides of the doors and windows and the corners of the walls are vertical lines. The top edges of the doors and windows and the ridgeline of the roof are horizontal lines. The sloping edges of the roof are inclined lines.

You recall from Chapter 1 that **parallel lines** are lines having the same direction. Notice that all the vertical lines in Figure 3-1 are parallel to each other and perpendicular to the horizontal lines, but that not all horizontal lines are parallel to each other. These relationships between vertical and horizontal lines are always true.

Figure 3-1 Find the vertical (V), horizontal (H), inclined (I), parallel, and perpendicular lines in this representation of a house.

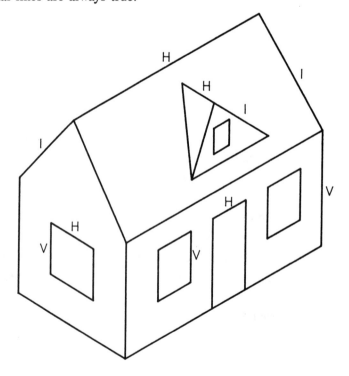

$$\angle X1 = \angle X2$$
$$\angle Y1 = \angle Y2$$

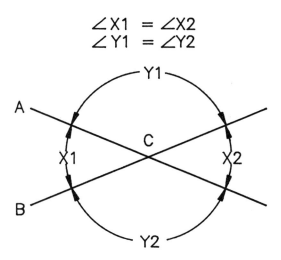

Figure 3-2 Lines A and B intersect at point C. Angle X1 is equal to angle X2 and angle Y1 is equal to angle Y2. C is the vertex.

Two lines that cross or touch each other are called **intersecting lines,** and together they form a set of **angles.** The point at which they meet, or intersect, is called the **vertex** of the angles and the lines are called the **sides.** The opposite angles formed by two intersecting straight lines, such as Y1 and Y2 and X1 and X2 in Figure 3-2, are always equal to each other. You recall from Chapter 2 that angles are measured by using a protractor, and that the units of measure are degrees. Two perpendicular intersecting lines form four angles of 90° each. These 90° angles are called **right angles** and are sometimes indicated by a small square drawn at the intersection. All angles less than (<) 90° are called **acute angles.** All angles greater than (>) 90°, but less than 180°, are called **obtuse angles.** A 180° angle is a straight line, as shown in Figure 3-3. Those angles greater than 180° are called **open angles.**

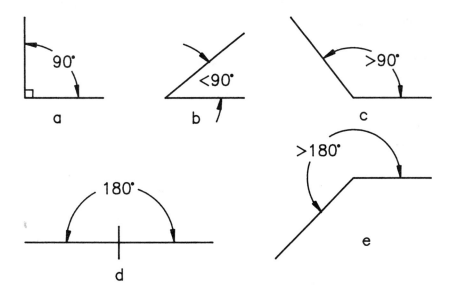

Figure 3-3 Types of angles: (a) right, (b) acute, (c) obtuse, (d) straight, and (e) open.

Straight-Line Constructions

This section discusses geometric construction methods involving straight lines. Refer to the cited figure as you read each description.

Drawing a line parallel to another line.   See Figure 3-4. With your compass set to match the required distance (R) between lines, draw two equal arcs with centers at any two different points on the given line AB. Lay your straightedge so that it just touches both arcs and draw line CD. Line CD is parallel to AB and distance R away from it.

Refer to Chapter 2, Section 3, to see how to draw one line parallel to another by using two triangles.

Drawing a line perpendicular to another line.   See Figure 3-5a, which shows how to draw a perpendicular line from a point off the given line. Begin at the given point, A, which is off the given line (BC). Using A as a center, draw an arc with radius R1 so the arc intersects the line BC at points 1 and 2. Using point 1 as a center, draw an arc with radius R2. Using point 2 as a center, draw an arc with the same radius (R2) so that the two arcs intersect at D. Connect points A and D with a straight line. Line AD is perpendicular to line BC.

Figure 3-5b shows how to draw a perpendicular line from a point on the line. Begin at the given point (E), which is on the given line (FG). Using E as a center, draw an arc with radius R1. The arc should intersect FG at points 3 and 4. Using points 3 and 4 as centers, draw two arcs with equal radii (R2). The arcs should intersect at point H. Draw a straight line through points E and H. Line EH is perpendicular to line FG.

Refer to Chapter 2, Section 2, to see how to draw one line perpendicular to another by using two triangles.

Dividing a straight line into any number of equal parts.   In this example you will learn to divide a line into seven equal parts. You can use the same method to divide a line into any number of equal parts that you wish.

Figure 3-4   Using a compass to draw lines CD parallel to and R distance from line AB.

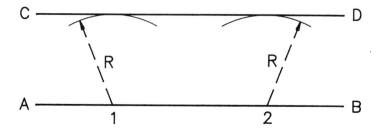

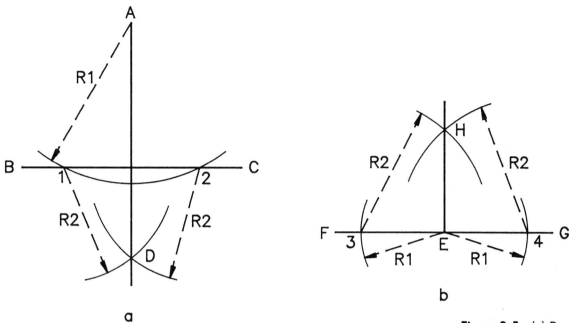

b

a

Figure 3-5 (a) Drawing a line perpendicular to line BC by using a point that does not lie on line BC. (b) Drawing a line perpendicular to line FG by using a point on line FG.

In Figure 3-6 the given line is line AB. Through one end of line AB, draw line BC perpendicular to AB. Place a scale so that the zero mark is at A and any seventh consecutive division intersects line BC. Mark the six intervening division points, and draw lines parallel to BC from the marked points to line AB. The resulting intersections on AB divide the line, in this example, into seven equal parts. The divisions used on the scale may be any that are convenient, but they must be equal to each other.

Figure 3-6 Dividing a line into seven equal parts.

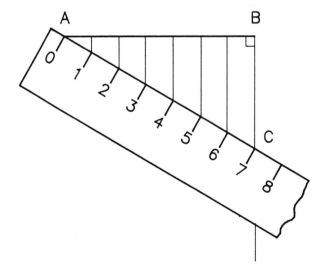

Bisecting a straight line.   **Bisect** means to divide in half. Figure 3-7 shows how to bisect line AB. Using points A and B as centers, draw any size arcs with equal radii (R1). The arcs should intersect at points C and D. Draw a straight line through points C and D. Line CD bisects and is perpendicular to line AB.

Bisecting an angle.   Figure 3-8 shows how to bisect the angle ABC. Using point B as a center, draw any size arc with radius R1 so that the arc intersects line AB at point D and line BC at point E. With D and E as centers, draw arcs with equal radii (R2) so that the arcs intersect at point F. Draw line BF, which bisects the angle ABC.

Transferring an angle.   To **transfer** an angle is to copy it. Figure 3-9 shows how to transfer angle ABC so that line AB lies along line DE. As shown in Figure 3-9a, use point B as a center, and draw an arc with any convenient radius, R1, so that the arc intersects lines AB and BC at points 1 and 2. Mark off points A1 and B1 on line DE in Figure 3-9b. Using point B1 as a center and a radius equal to R1, draw the arc with radius R2. The arc should intersect DE at point 3. Using point 3 as a center and the distance from point 1 to point 2 as the radius R3, draw an arc that passes through point 4. Draw a line from B1 through point 4 and mark off point C1. The new angle, A1B1C1, is the same size as the angle ABC.

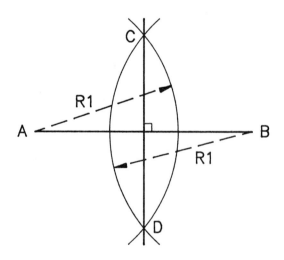

Figure 3-7   Bisecting line AB.

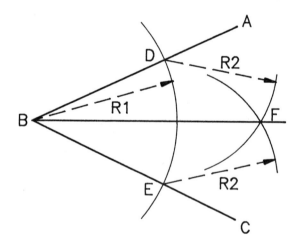

Figure 3-8   Bisecting angle ABC.

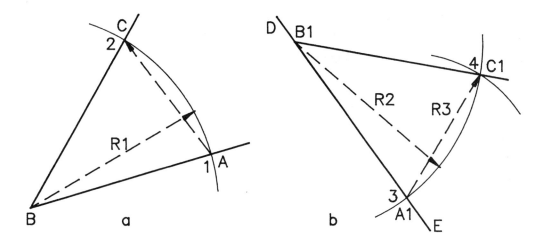

Figure 3-9 Transferring angle ABC, shown in part a, so that line AB lies along line DE, as shown in part b.

3-3 THE GEOMETRY OF CURVED LINES

Definitions

There are only two basic kinds of lines, straight and curved, so any line that is not straight is a **curved line.** A curved line's direction is constantly changing, and the way it changes determines whether the line is a definable curve, such as a circle, or an irregular curve.

Figure 3-10 illustrates the most important curved lines, which are circles. **Circles** are lines that begin and end at the same point, and are everywhere the same distance from another point called the **center.** In other words, if you measured from the center to any points on a given circle, the distances measured would be equal. This measurement is called the **radius,** and it is the distance you set your compass to when using one to draw a circle. However,

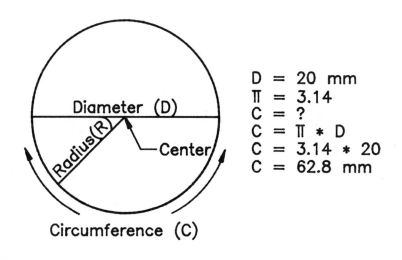

Figure 3-10 The principal parts of a circle—center, circumference, diameter, and radius—and calculating the length of the circumference.

$$D = 20 \text{ mm}$$
$$\pi = 3.14$$
$$C = ?$$
$$C = \pi * D$$
$$C = 3.14 * 20$$
$$C = 62.8 \text{ mm}$$

a circle is most commonly identified by its **diameter,** which is twice the radius. The linear distance around the circle is called the **circumference.** This distance is usually calculated, because measuring a curved line directly is difficult. The formula used to calculate the length of a circumference is

$$C = \pi * D \text{ or } C = \pi * 2R$$

in which

> C = circumference
>
> π = 3.14
>
> D = diameter
>
> R = radius

A portion of a circle is called an **arc.** An arc is specified by its radius and included angle. Figure 3-11 shows an example of an arc and its included angle, angle A. The arc is said to **subtend** the angle. The length of an arc depends on its radius and the size of its angle. The larger either of them becomes, the longer the arc is. Since a circle contains an angle of 360°, you can calculate the length of an arc by using the formula

$$L = \pi * 2R * (A/360)$$

in which

> L = length of arc
>
> π = 3.14
>
> R = radius
>
> A = angle

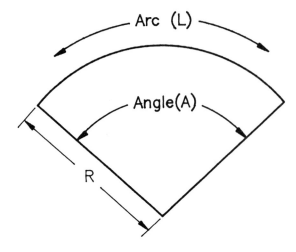

R = 30 mm
A = 120°
π = 3.14
L = ?
L = π * 2R * (A/360)
L = 3.14 * 60 * 120/360
L = 62.8 mm

Figure 3-11 The arc of a circle, its radius, and its included angle and calculating the length of the arc.

A straight line that passes through a circle and intersects it at two different points forms a **chord** of the circle. A straight line that touches a circle at only one point is called a **tangent** to the circle. A curved line may also be tangent to a circle. Both the perpendicular bisector of a chord and a perpendicular line drawn from the point of tangency pass through the center of the circle, as shown in Figure 3-12.

Two circles of different sizes, with the same center, are called **concentric circles.** Two circles of different sizes and different centers are called **eccentric circles.** Figure 3-13 shows both types.

Another important type of curved line is the **ellipse.** The ellipse can be defined in several different ways, but it is easiest to visualize it as a circle rotated about one of its diameters. A circle appears round only if you view it straight on; otherwise, it appears foreshortened in one direction, as Figure 3-14 illustrates. An ellipse has two axes. The major axis is equal in length to

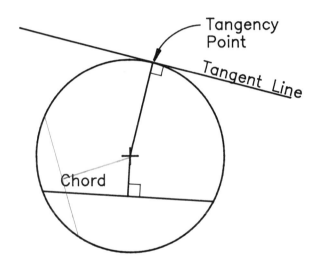

Tangency Point

Tangent Line

Chord

Figure 3-12 A tangent line, tangent point, and chord of a circle.

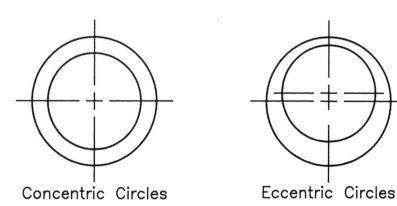

Concentric Circles

Eccentric Circles

Figure 3-13 Concentric and eccentric circles.

Figure 3-14 An ellipse developed by rotating a circle about its major axis.

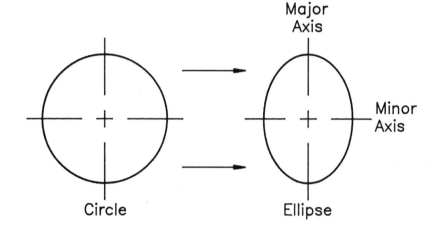

the diameter of the original circle; the minor axis length depends on the degree of rotation. The farther you rotate, or tilt, a circle away from you, the shorter the minor axis becomes. The major axis, however, always remains equal to the diameter of the original circle. Note that, though the circle diameter may appear shorter, it can never appear longer than its actual length.

Curved-Line Constructions

Descriptions of geometric construction methods involving curved lines follow. Refer to the cited figure as you read each description.

Bisecting an arc. See Figure 3-15. Using A and B as centers, draw any

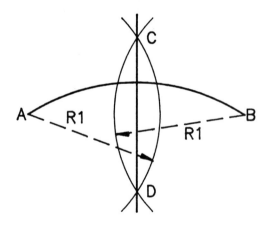

Figure 3-15 Bisecting an arc.

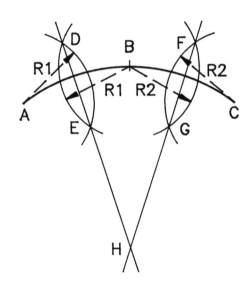

Figure 3-16 Drawing an arc passing through three points.

arcs with equal radii (R1). The arcs should intersect at C and D. Draw a straight line through points C and D. Line CD bisects arc AB.

Drawing an arc passing through three points.   In Figure 3-16 the three given points are A, B, and C. Using A and B as centers, draw arcs with equal radii (R1). The arcs should intersect at points D and E. With B and C as centers, draw arcs with equal radii (R2). The arcs should intersect at points F and G. Lines DE and FG intersect at point H. Using H as a center, draw an arc through A, B, and C.

Drawing an arc tangent to a right angle.   Figure 3-17 shows lines AB and AC, which form a right angle. R is the given radius for the arc. With point A as the center and radius R, draw an arc intersecting the lines at D and E. With D and E as the centers and radius R, draw arcs intersecting at F. Using F as the center and R as the radius, draw the arc that is tangent to the straight lines at points D and E.

Drawing an arc tangent to any two nonparallel straight lines.   In Figure 3-18a AB and CD are two given lines that form an obtuse angle and R is the given radius for the arc. From any point on AB, draw an arc with R as the radius. From any point on CD, draw an arc with R as the radius. Tangent to these arcs, draw two straight lines, one parallel to AB and one parallel to CD. The two new lines should intersect at point E. Using E as the center and R as the radius, draw the required arc. Perpendiculars from E to AB and CD determine the exact points of tangency at 1 and 2. Figure 3-18b shows the same process used with lines that form an acute angle.

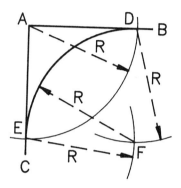

Figure 3-17 Drawing an arc tangent to a right angle.

Figure 3-18 (a) Drawing an arc tangent to two non-parallel straight lines that form an obtuse angle. (b) Drawing an arc tangent to two nonparallel straight lines that form an acute angle.

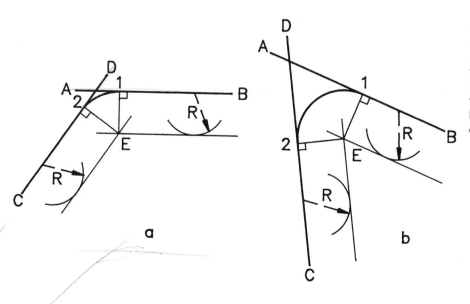

Figure 3-19 (a) Drawing
an arc tangent to a straight
line and the inside of an arc.
(b) Drawing an arc tangent
to a straight line and the
outside of an arc.

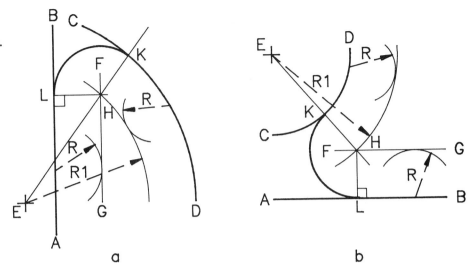

a   b

Drawing an arc tangent to a straight line and an arc. In Figure
3-19a AB is the given straight line, CD is the given arc with E as its center,
and R is the radius of the arc to be drawn. Draw straight line FG parallel to
line AB and at distance R from it. With E as the center, draw the arc with
R1 as its radius. The arc should be concentric with, and at distance R from,
arc CD. This arc should intersect line FG at point H. Using H as the center
and R as the radius, draw the required arc. K and L are the points of tangency.

Figure 3-19b shows how to use the same process to draw an arc tangent
to a straight line and a reversed arc.

Drawing a concave arc tangent to two arcs. In Figure 3-20 AB and
CD are the given arcs. The arc centers are E and F, respectively. R is the given
radius for the arc to be drawn. Using E and F as centers, draw two arcs with
the radii R1 and R2. The two arcs should be concentric with the two given
arcs at distance R from them and intersecting at point G. Using G as the center
and R as the radius, draw the required arc. K and L are the points of tangency.

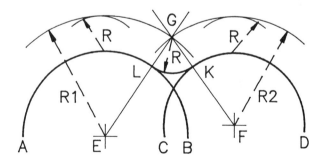

Figure 3-20 Drawing a concave arc tangent to two arcs.

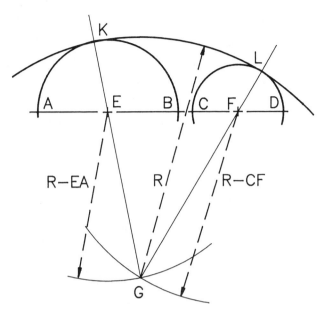

Figure 3-21 Drawing a convex arc tangent to two arcs.

Drawing a convex arc tangent to two arcs.   In Figure 3-21 AB and CD are the given arcs. The arc centers are E and F, respectively. R is the given radius for the arc to be drawn. Using E and F as centers, draw two arcs: one equal to R − EA and one equal to R − CF. The arcs should intersect at point G. Using G as the center and R as the radius, draw the required arc. K and L are the points of tangency.

Drawing a reverse curve connecting two parallel lines.   Figure 3-22 shows how to draw a reverse, or ogee, curve. In the figure AB and CD are the given parallel lines. Draw perpendiculars at points B and C. Draw line BC and assume point G where you desire the curve to reverse. The perpendicular bisectors of BG and GC (lines 1 and 2) intersect the perpendiculars from B and C at points E and F. Using E as the center and radius BE, draw arc BG. Using point F as the center and radius FC, draw arc GC.

Figure 3-22 Drawing a reverse curve connecting two parallel lines.

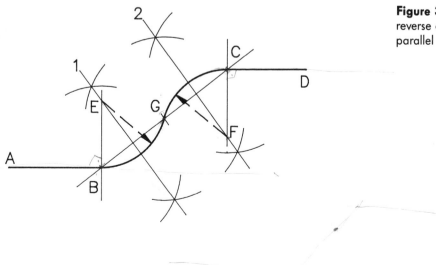

Drawing an ellipse by the trammel method. Figure 3-23 shows the major axis AC perpendicular to the minor axis BD. Make a trammel by marking lengths E1A1 and E1D1 on the straight edge of a strip of paper. These lengths are equal to one-half of each axis. Place your trammel so that point D1 is on the major axis and point A1 is on the minor axis. A pencil mark at E1 defines one point on the curve of the ellipse. Mark similar points by moving the trammel, being careful at all times to keep point A1 on the minor axis and point D1 on the major axis. Plot enough points to establish the shape of the ellipse, and connect them by using the irregular curve to complete the figure. Remember, the more points you plot, the more accurate your ellipse will be.

Drawing an approximate ellipse. Figure 3-24 shows the major axis AB and the minor axis CD. Draw diagonal line AD. With E as the center and radius EA, draw arc AF. With D as the center and radius DF, draw arc FG. Draw the perpendicular bisector, HJ, of line AG. Points K and J are used as centers for drawing circular arcs to create one half of the ellipse. Mark the other two centers, L and M, by setting off distances from point E equal to EJ and EK, respectively. Draw diagonal lines as shown to determine the meeting points (1, 2, 3, and 4) of the arcs.

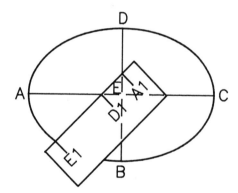

Figure 3-23 Drawing an ellipse by the trammel method.

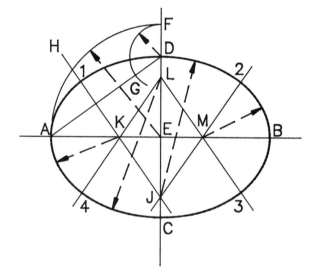

Figure 3-24 Drawing an approximate ellipse.

3-4 THE GEOMETRY OF PLANE FIGURES

Definitions

Plane figures are those geometric shapes that are flat and have two dimensions. Plane figures are classified according to their shape, number of sides, and the relationship between those sides. Plane figures with one or two sides must be bounded by curved lines. You have already learned about the most common of these figures, circles and ellipses. Plane figures bounded by straight sides are called **polygons.**

Three-sided polygons are called **triangles.** Triangles have three interior angles that always contain a total of 180°, no matter what the configuration or size of the triangle. The most important types are **right triangles** (they contain one 90° angle), **isosceles triangles** (they contain two equal sides and angles), and **equilateral triangles** (they contain three equal sides and angles). These are illustrated in Figure 3-25.

Because plane figures are two-dimensional, they are measured by their **areas.** Some common units of measurement for area are square millimeters (mm^2), square inches (in.^{2}), square meters (m^2), and square feet (ft^2). A very common unit of area is the acre, used for measuring land or property. The area of a circle (refer to Figure 3-10) is calculated by the formula

$$A = \pi * R^2$$

in which

A = area

π = 3.14

R^2 = radius squared

Figure 3-25 Three important types of triangles.

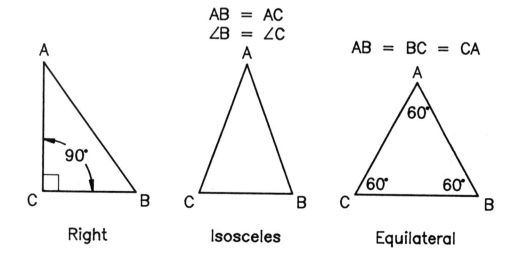

AB = AC
∠B = ∠C

AB = BC = CA

Right   Isosceles   Equilateral

For example:

R = 5 ft

$A = \pi * 5^2$

A = 3.14 * 5 * 5

$A = 78.5 \text{ ft}^2$

The area of a triangle can be calculated by using the formula:

A = 0.5b * h

in which

A   = area

b   = base length (any side)

h   = height (measured perpendicular to the base)

Figure 3-26 illustrates the use of this formula.

Four-sided polygons are called **quadrilaterals.** Several types are of particular importance. A **parallelogram** is any quadrilateral that has two sets of parallel sides. If the sides are also equal in length, it is called a **rhombus.** Any four-sided polygon that has only two parallel sides is called a **trapezoid.** A **rectangle** is a parallelogram that has four 90° interior angles, and a **square** is a rectangle with four equal-length sides. Figures 3-27 and 3-28 show examples of four-sided polygons.

To calculate the area of any parallelogram, use the formula

A = b * h

in which

A   = area

b   = base length (any side)

h   = height (measured perpendicular to the base)

Figure 3-29 illustrates the use of this formula. Note that this formula works only for parallelograms—not for other quadrilaterals, such as trapezoids.

Figure 3-26 Calculating the area of a triangle.

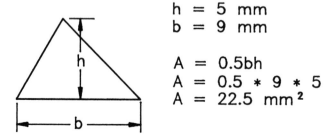

h = 5 mm
b = 9 mm

A = 0.5bh
A = 0.5 * 9 * 5
$A = 22.5 \text{ mm}^2$

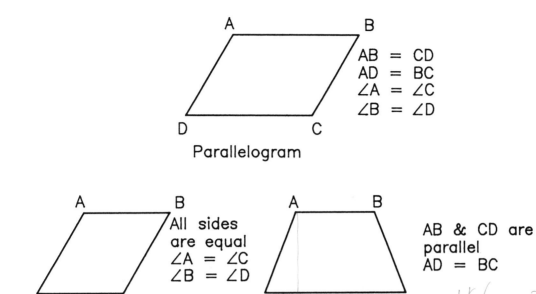

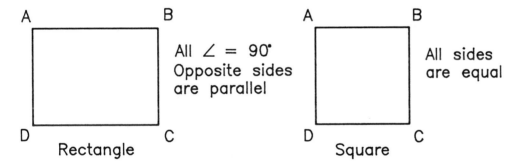

Figure 3-27   The parallelogram, rhombus, and trapezoid.

All ∠ = 90°
Opposite sides
are parallel

All sides
are equal

Rectangle   Square

Figure 3-28   The rectangle and the square.

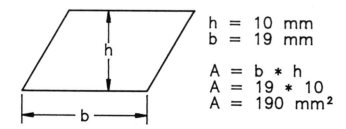

h = 10 mm
b = 19 mm

A = b * h
A = 19 * 10
A = 190 mm²

Figure 3-29   Calculating the area of a parallelogram.

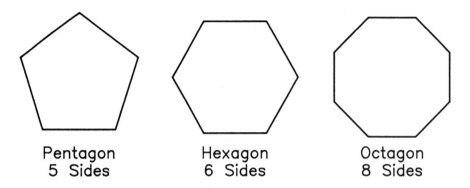

Figure 3-30 Regular polygons.

Other polygons are named by the number of sides they have: the **pentagon** has five sides, the **hexagon** has six sides, and the **octagon** has eight sides. **Regular polygons** are figures that have equal sides and angles. Thus, equilateral triangles and squares are regular polygons. Figure 3-30 shows examples of regular polygons with more than four sides.

A regular polygon can have a circle drawn inside that touches each side so that there are a maximum number of tangency points (one for each line in the figure). This circle is called an **inscribed** circle. A **circumscribed** circle is drawn outside the polygon in such a way that it touches the vertices of each angle. See Figure 3-31 for examples.

Since all quadrilaterals have four sides and four angles, it may have occurred to you that all polygons will have an equal number of sides and angles. In other words, each side will have a corresponding angle opposite it. A fixed number of total degrees is contained in each different straight-sided figure, and the total becomes larger as the number of sides and angles increases. It would be difficult to remember what all those totals are, but fortunately there is a simple formula to calculate them.

Figure 3-31 (a) A circle inscribed in a square. (b) A circle circumscribed around a hexagon.

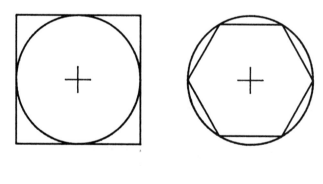

a   b

degrees = (N − 2) ∗ 180°

in which

degrees = total for any polygon

N   = number of sides

Using the formula for a triangle:

$$
\begin{aligned}
\text{degrees} &= (N - 2) * 180° \\
&= (3 - 2) * 180° \\
&= 1 * 180° \\
&= 180°
\end{aligned}
$$

Using the formula for a quadrilateral:

$$
\begin{aligned}
\text{degrees} &= (N - 2) * 180° \\
&= (4 - 2) * 180° \\
&= 2 * 180° \\
&= 360°
\end{aligned}
$$

From these results you can see that the sum of all the interior angles in any triangle equals 180° and in any quadrilateral, 360°. A pentagon contains 540° and a hexagon 720°. By comparing these totals (180-360-540-720), it is readily apparent that the difference between each one is 180°. In others words, each additional side and angle increases the total angular measurement of all angles by 180°.

Plane-Figure Constructions

Descriptions of geometric construction methods involving plane figures follow. Refer to the cited figure as you read each description.

Drawing a triangle. In Figure 3-32a, A, B, and C are three given sides. Draw line RS equal in length to given side C. Using point R as the center and side B as the radius, draw an arc. With S as the center and side A as the radius, draw a second arc that intersects the first arc at T. The triangle RST is identical to triangle ABC.

Figure 3-32b shows that a right triangle will be constructed if the sides have lengths of 3, 4, and 5 units.

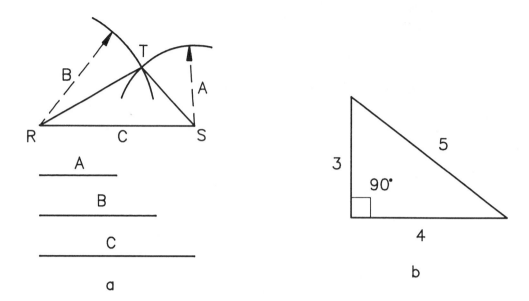

Figure 3-32 (a) Drawing a triangle with sides equal to A, B, and C. (b) A triangle with sides of 3, 4, and 5 units is a right triangle.

Drawing a square. In Figure 3-33a, side AB is the given side. Your goal is to draw a square that includes AB. Begin by drawing a circle with a radius equal to one-half AB (AB/2). By any method, draw two sets of perpendicular lines tangent to the circle to form a square.

In Figure 3-33b, AC is the given distance. In this case, AC is specified as the diagonal of the square you will draw. Draw lines through A and C at angles of 45° to line AC so that they intersect each other at B and D.

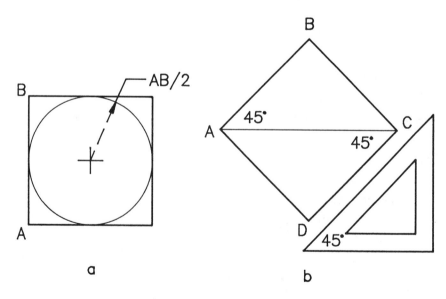

Figure 3-33 (a) Drawing a circle with the radius AB/2. (b) Drawing a square with the diagonal AC.

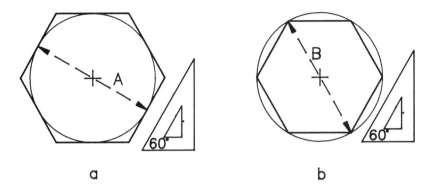

Figure 3-34 (a) Drawing a regular hexagon when the short diameter is given. (b) Drawing a regular hexagon when the long diameter is given.

Drawing a regular hexagon.   Your first goal is to draw a regular hexagon when only the short diameter—that is, the distance between the flat sides of the hexagon—is given. In Figure 3-34a, the short diameter is given as distance A. Draw a circle with a diameter equal to A. With a 30°-60° triangle, draw sides tangent to the circle, as shown.

Your next goal is to draw a regular hexagon when only the long diameter—that is, the distance between opposite corners of the hexagon—is given. In Figure 3-34b, the long diameter is given as distance B. Draw a circle with a diameter equal to B. With a 30°-60° triangle, draw the sides within the circle as shown. Note: The length of each side of the hexagon is equal to the radius of the circumscribed circle.

Drawing a regular octagon.   In Figure 3-35 only the short diameter, D, of a regular octagon is given. To create the octagon, draw a circle with a diameter equal to D. With a 45° triangle, draw sides tangent to the circle, as shown in the figure.

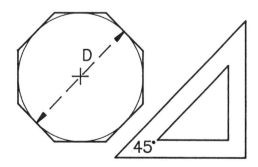

Figure 3-35   Drawing a regular octagon when the short diameter is given.

Figure 3-36 Drawing a pentagon inside a circle.

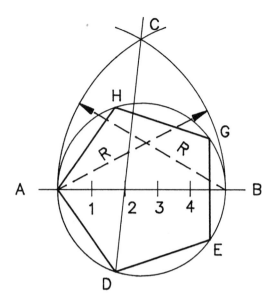

Drawing a regular polygon (of any number of sides). In Figure 3-36 the only given is a circumscribing circle. To create a pentagon, divide the diameter AB into as many equal parts as the polygon has sides (five in this example). With R as the radius and A as the center, draw an arc. With B as center and R as the radius, draw another arc. The two arcs should intersect at point C. Draw a line from C through point 2, intersecting the circle at D. Draw line AD, which is one side of the polygon. Mark off distance AD around the circle by using a compass, and draw the remaining sides.

Transferring plane figures. In Figure 3-37a the given polygon is drawn with thick lines. To transfer it, draw a rectangle around the given polygon. Draw a rectangle of the same size in the new location, and locate the corners (vertices) of the polygon by transferring distances from the corners of the circumscribed rectangle.

Figure 3-37b shows a figure that includes irregular curves. To transfer it, construct a rectangle around the figure, and draw offset grid lines to intersect

Figure 3-37 (a) Transferring a given polygon. (b) Transferring a figure that includes irregular curves.

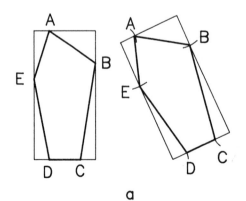

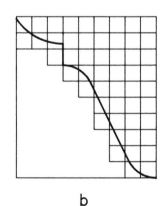

a   b

the outline at a number of defining points. In the new location construct the rectangle, as you did in part a, but in this case include the grid lines. Transfer the defining points and draw a curve through the points to complete the new figure.

3-5 THE GEOMETRY OF SOLID FIGURES

Solid figures are three-dimensional. They can be described as plane figures with depth. Their principal measurement is **volume.** The measurement units for volume are length units cubed—cubic millimeters (mm^3) and cubic inches (in.^{3}), for example. Of the many possible solids, prisms, cubes, pyramids, cones, cylinders, spheres, and toruses are of particular importance. All these solid figures are illustrated in Figure 3-38.

Solids bounded by plane (flat) surfaces are called polyhedra. (The singular form of *polyhedra* is **polyhedron.**) **Prisms** are polyhedra that have two equal, parallel bases and three or more lateral faces. **Cubes** are the best-known examples of polyhedra. **Pyramids** have polygonal bases and three or more triangular faces that meet at a common point called the **vertex,** or **apex.** Solids with circular or elliptical bases and single curved lateral surfaces that extend to a vertex are called **cones. Cylinders** have two circular or elliptical bases and single curved, lateral surfaces. Both cones and cylinders have lateral surfaces that are **single curved**—that is, they curve in only one direction. On the other hand, **spheres** and **toruses** have single surfaces that are **double curved**—that is, they curve in two directions. The surface of a sphere is everywhere the same distance (radius) from its center. A torus is doughnut-shaped.

Figure 3-38 The principal solid geometric forms.

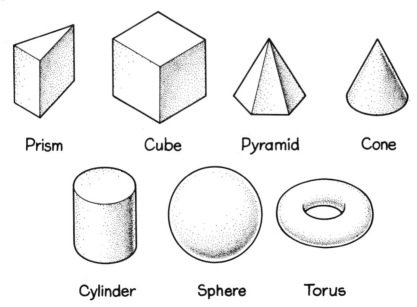

Prism   Cube   Pyramid   Cone

Cylinder   Sphere   Torus

3-6 CONCLUSION

This chapter has covered some basic concepts of the geometry of plane and solid figures. The chapter presented definitions of straight, curved, vertical, horizontal, parallel, perpendicular, and inclined lines. Also discussed were intersecting lines and angles and tangencies between straight and curved lines. The chapter introduced plane figures (including circles and ellipses) and polygons (such as triangles, squares, and rectangles). You learned some basic formulas for calculating various sizes. Construction techniques for solving various problems connected with all these were presented. Lastly the chapter defined and illustrated solid geometric figures.

To fully understand many of the concepts that will be presented in the remainder of this book, you will need to understand these definitions. Mastering the construction techniques will help improve your drawing skills and give you another graphical problem-solving skill.

Check your general understanding of this chapter by answering the following questions.

REVIEW QUESTIONS

1. Which of the following drafting instruments is *not* necessary for doing geometric constructions?
    a. Straightedges (triangles)
    b. compass
    c. scale
    d. ellipse template

2. Equal and opposite angles are formed when two straight lines are:
    a. parallel
    b. perpendicular
    c. intersecting
    d. tangent

3. A bisector divides _____ in half.
    a. a straight line
    b. a curved line
    c. an angle
    d. all the above

4. What is the circumference of a circle with a 50-mm diameter?

   a. 31.4 mm
   b. 314.1 mm
   c. 78.5 mm
   d. 785.0 mm

5. What is the length of a 30° arc with a radius of 75 mm?

   a. 39.3 mm
   b. 19.6 mm
   c. 22.7 mm
   d. 31.3 mm

6. Tangency between two lines requires that

   a. both lines be straight
   b. both lines be curved
   c. at least one line be straight
   d. at least one line be curved

7. Two different-sized circles with different centers are called:

   a. elliptical
   b. concentric
   c. eccentric
   d. circumscribed

8. The fact that the perpendicular bisector of the chord of a circle passes through the center of that circle is particularly useful in drawing:

   a. an arc tangent to two straight lines
   b. a bisector of an angle
   c. a circle through three points not in a line
   d. a concentric circle ellipse
   e. both b and d

9. The center of an arc with a 50-mm radius tangent to two lines is _____ away from each line.

   a. 100 mm
   b. 50 mm
   c. 25 mm
   d. can't say from information given

10. What is the area of an equilateral triangle with sides of 10 in. and a height of 8.66 in.?

    a. 40.3 in.^{2}
    b. 21.65 in.^{2}
    c. 86.6 in.^{2}
    d. 43.3 in.^{2}

11. How many total degrees are there in the interior angles of an octagon?

  a. 720°

  b. 960°

  c. 1080°

  d. 1260°

$(8-2)(180)$
6×180
1080

12. Which of the following is *not* a solid figure?

  a. prism

  b. cylinder

  c. torus

  d. parallelogram

PROBLEMS

For problems 3-1 through 3-17, trace each problem onto vellum or transfer it to vellum by means of measurements.

3-1 Divide line AB into seven equal parts.

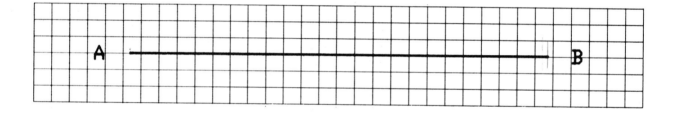

3-2 Construct a perpendicular bisector of line CD.

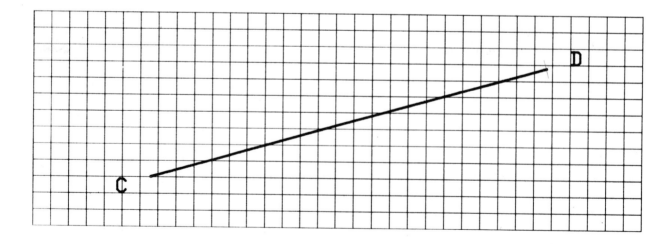

3-1   3 2
3-3   3-4

3-3   Construct lines from P and Q that are perpendicular to line MN.

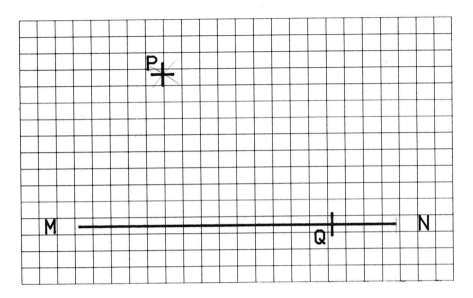

3-4   Construct the bisector of angle XYZ.

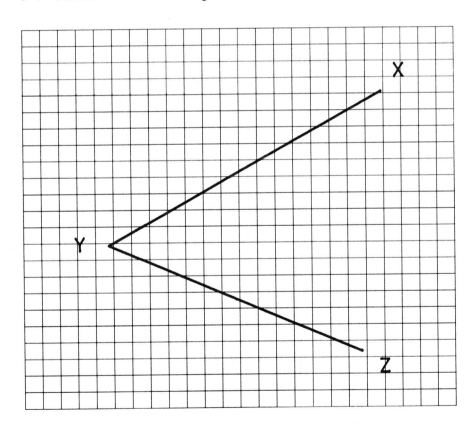

3-5   Construct the bisector of arc CD and mark the center of arc CD.

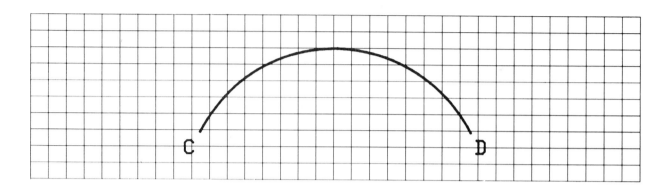

3-6   Reconstruct angle ABC so that points B and C lie on line 1,2.

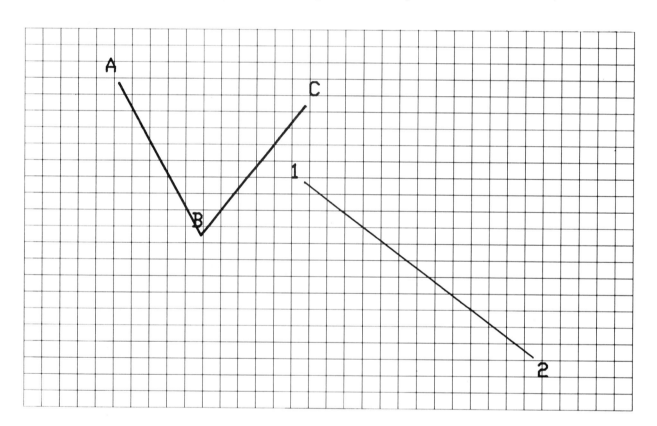

3-7   Draw a 1-in.–radius arc tangent to the two given lines, and mark the tangent
points.

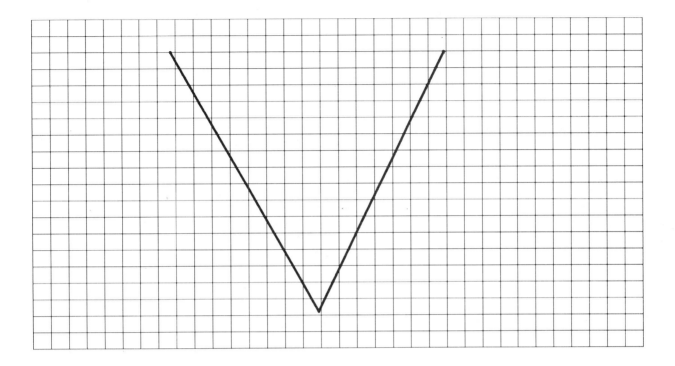

3-8   Draw a 45-mm–radius arc tangent to the two given lines, and mark the tangent
points.

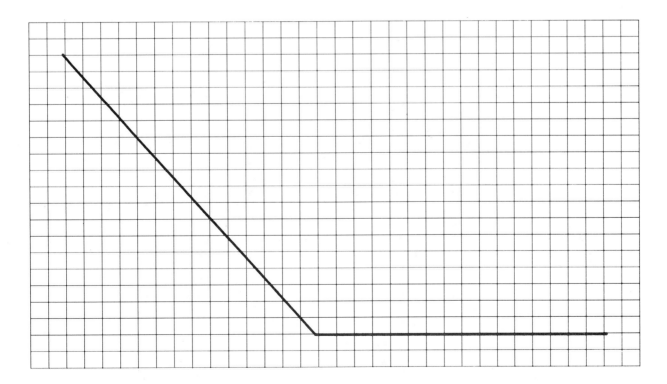

3-9 Draw a 33-mm–radius arc tangent to the given arc and line. Mark the tangent points. Hint: Find the center of the given arc first.

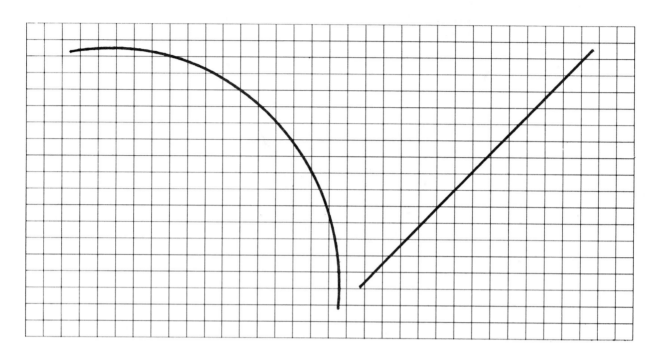

3-10 Draw a .75-in–radius concave arc and a 3.50-in.–radius convex arc tangent to the two given arcs. Mark the tangent points. Arc 1 has a 1.50-in. radius and arc 2 has a 1.00-in. radius.

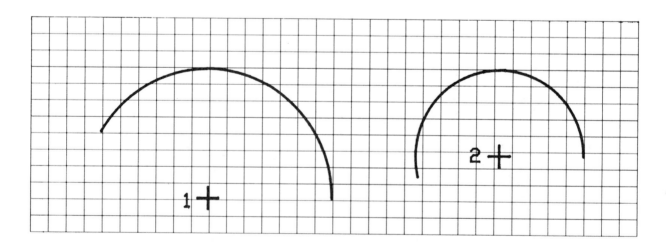

3-11   Draw a reverse, or ogee, curve with equal radii tangent to the two given lines.

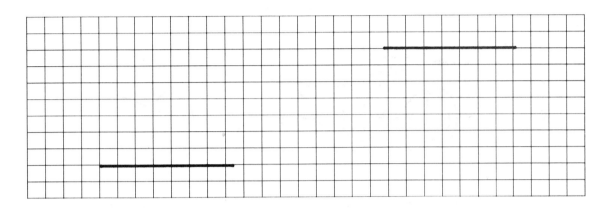

3-12   Measure the angle subtended by arc EF and calculate the length of the arc.

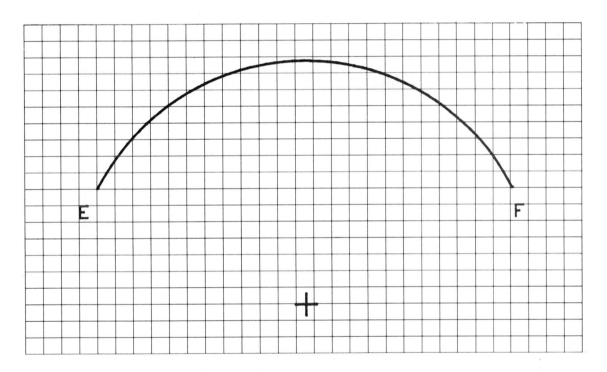

3-13   Inscribe a polygon with seven equal sides in a circle with a diameter of 70 mm.

3-14   Inscribe a square in a circle, which has a 3-in. diameter. The sides of the square should be parallel to the centerlines.

3-15 Redraw the figure shown to scale. Each polygon should be inscribed or circumscribed by a circle that has a 1.00-in. diameter. Polygons are located on centerlines spaced 1.50 in. apart.

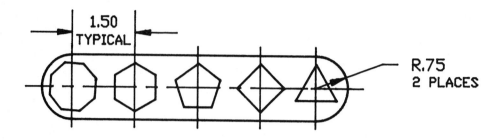

3-16 Construct a triangle with sides of the three lengths shown. Calculate its area and measure each of the angles.

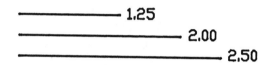

3-17 Construct three ellipses by using the major and minor axes given.

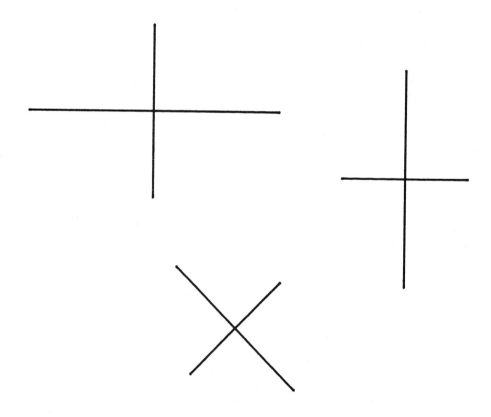

3-18 Construct figures a, b, and c full size on separate sheets of paper.

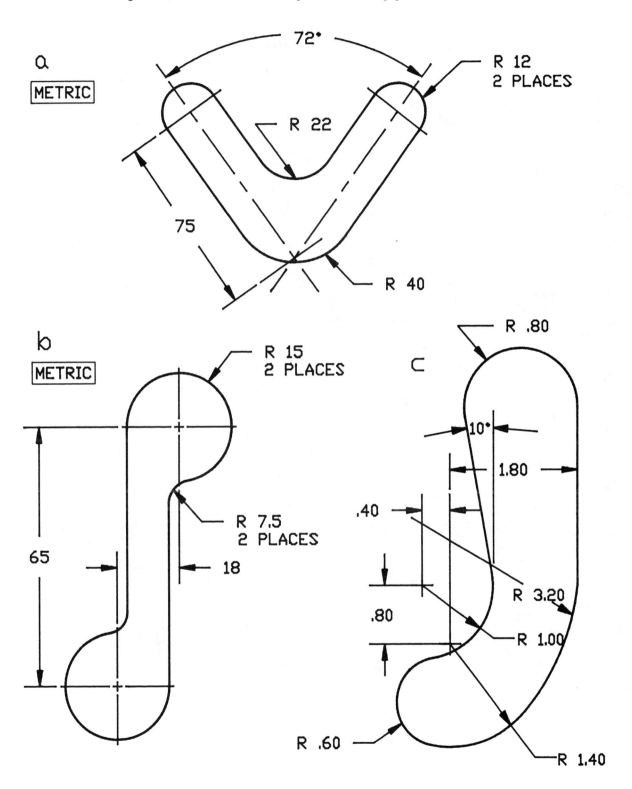

PICTORIAL DRAWINGS

4-1 INTRODUCTION

The illustration that opens this chapter is a pictorial drawing by the artist M. C. Escher. Have you ever looked at realistic drawings and paintings such as this and wondered how the artists made them? How do they achieve that sense of depth, that three-dimensional appearance on a two-dimensional piece of paper? They do it through the use of pictorial drawing, which is the subject of this chapter. Pictorial drawings are single-view drawings whose purpose is to show objects in a familiar and easily understandable way. They show objects in much the same way they appear in photographs.

All objects, no matter their size or shape, are three-dimensional. That is, they have width, height, and depth. But a drawing surface (the paper) has only two dimensions: width and height. As a result, pictorial drawings have distortions of size and shape that limit their usefulness as technical drawings. They do not show angles and surfaces in true size, but rather as they appear to be when viewed from a given direction. Because pictorial drawings show three sides of an object at one time, accuracy of dimension is sacrificed in favor of appearance. However, pictorial drawings require no special skill for interpretation or understanding. Thus they may be used to communicate technical information to people who are without technical training. Pictorial drawings are also useful as design or idea sketches and as aids in visualizing complex parts and assemblies.

Three different types of pictorial drawings are commonly used: perspective, oblique, and isometric. Examples are shown in Figure 4-1. Perspective drawings are the most realistic but also the most difficult to draw accurately. They are used primarily by architects and artists. Escher's drawing at the

After you complete this chapter, you should:

1. know the essential features of isometric, oblique, and perspective drawings and recognize the differences between them

2. be able to select the most appropriate type of pictorial to draw for any object

3. be able to draw clear and accurate pictorial drawings of objects with both straight and curved lines

4. be able to use shading techniques when appropriate

109

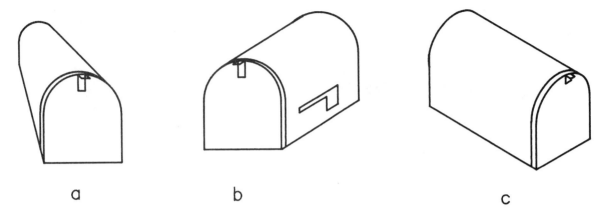

a   b   c

Figure 4-1   Three different pictorial drawings of a mailbox: (a) perspective, (b) oblique, and (c) isometric.

beginning of this chapter is a perspective, but it has something unusual about it. Can you identify it? Oblique drawings are the easiest to draw but are the least realistic. For some types of objects, however, they have certain advantages that make them quite useful. Isometric drawings are the most widely used pictorials in technical drawing because they are a compromise between realism and drawing difficulty. This chapter will discuss each of these three types separately, beginning with isometric drawing.

4-2 ISOMETRIC DRAWINGS

Isometric drawings are part of a class of drawings called axonometric—a class that also includes dimetric and trimetric drawings. One advantage axonometrics have over other pictorials is that they can be projected from multiviews because both types of drawings belong to a larger class called orthographic. Axonometrics, when projected from a multiview, are called **axonometric projections.** However, the projection process is complex and time-consuming, and you will ordinarily draw isometric drawings. These are not true projections, but they are much easier to draw.

Straight Lines in Isometric Drawings

Isometric drawings are begun by drawing three lines called **isometric axes.** These lines are drawn from a single corner point, radiating outward and upward with 120° angles between them, as shown in Figure 4-2a. These lines, and any others parallel to them, are called **isometric lines.** They are drawn full-size, and their true lengths can be measured. A simple cube constructed

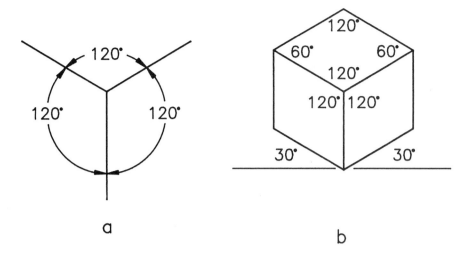

Figure 4-2 (a) The isometric axes and (b) an isometric cube.

of isometric lines only is shown in Figure 4-2b. Notice that the angles between all lines are either 60° or 120°. A more complex object constructed of isometric lines is shown in three different positions in Figure 4-3. All the lines are parallel to one of the axes; therefore, all lines used in the drawing are isometric lines.

All lines that are not parallel to one of the isometric axes are called **nonisometric lines**. They do not appear true length, so the measurement of nonisometric lines on an isometric drawing does not reflect the real measurement of the corresponding part of the object. Nonisometric lines are drawn by placing their end points on isometric lines and connecting them. Thus, all

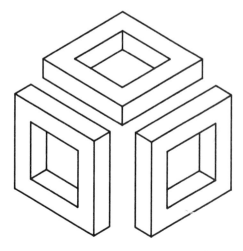

Figure 4-3 An object drawn in three different positions. All lines are isometric lines.

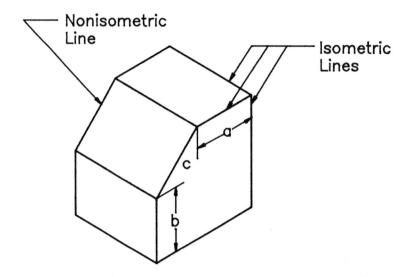

irregularly shaped objects are drawn starting with a box of isometric lines on which are located the **coordinates** of nonisometric lines, as shown in Figure 4-4.

This system, called **box construction,** or **coordinate construction,** eliminates the need for measuring angles on an isometric drawing. This is fortunate, because isometric angles never appear true size; they always appear larger or smaller than they actually are. You have already seen in Figure 4-2 that the angles between the isometric lines appear as 60° or 120°, but on a real cube these angles are all actually 90°. In Figure 4-4, the angle between the vertical axis and the nonisometric line C is 30°, but you cannot measure it on the drawing. However, you can measure the distances a and b along the isometric lines, thus establishing the end points of line c. This allows you to

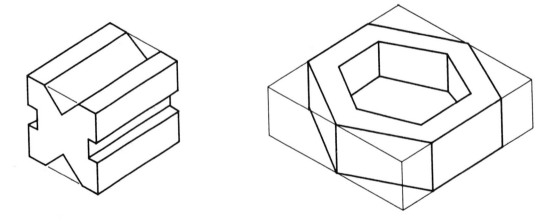

Figure 4-5   Isometric drawings of objects with nonisometric lines.

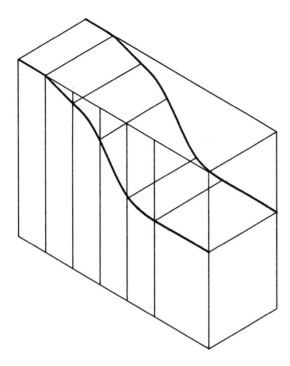

Figure 4-6 Drawing irregular curves by plotting points on isometric lines.

construct the line, without measuring its angle. More complex objects constructed in the same way are shown in Figure 4-5.

Curved lines are drawn by a method similar to that for nonisometric lines. However, a number of points must be placed on isometric lines and connected by an irregular curve, as shown in Figure 4-6.

Circles in Isometric Drawings

Since no angles and only some lines appear true size on an isometric drawing, it follows that surfaces never appear true size and shape. It also follows that circles do not appear true size and shape, but rather as ellipses. An ellipse is most easily drawn using an ellipse template, but if one is not available, you must construct the ellipse. The three steps in the **four-center technique** for drawing approximate ellipses are described in the list that follows. Figure 4-7a illustrates the steps.

1. Draw an isometric square (a rhombus) whose sides are the same length as the diameter of the original circle and whose center is located at the desired ellipse's center.

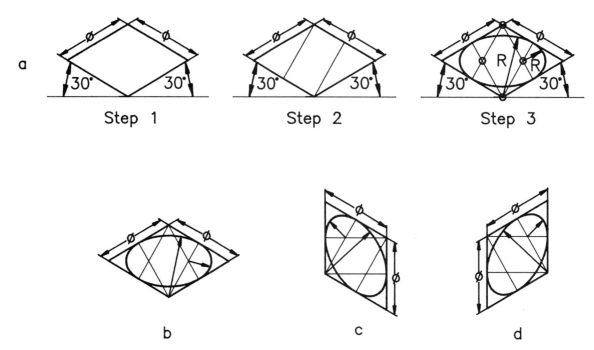

Figure 4-7 The steps in constructing ellipses by using the four-center technique.

2. Draw diagonal lines from the vertices of the two larger angles to the midpoints of the opposite sides.

3. The vertices of the two larger angles and the points where the diagonals cross are the four centers for drawing four arcs by using a compass or circle template.

Figure 4-7b shows a completed ellipse drawn on a top plane; Figure 4-7, parts c and d, show the ellipse drawn on side planes.

Cylinders are drawn by constructing the two elliptical ends and then drawing lines parallel to the centerline of the cylinder and tangent to the ellipses (see Figure 4-8). Holes are drawn in a similar manner, as shown in Figure 4-9.

Arcs are represented as portions of ellipses. Arcs are constructed by the four-center method, omitting any unneeded portions, as shown in Figure 4-10.

The Box Construction Method

Isometric drawings are most easily drawn by using the box construction method. In this method a rectangular box, which just encloses the object, is drawn with light construction lines. Next, the major features, which show the

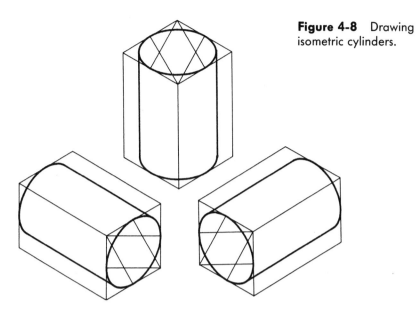

Figure 4-8 Drawing isometric cylinders.

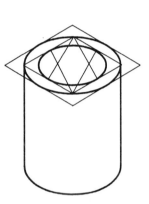

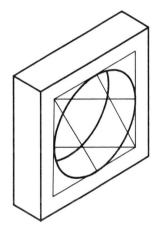

Figure 4-9 Drawing isometric holes.

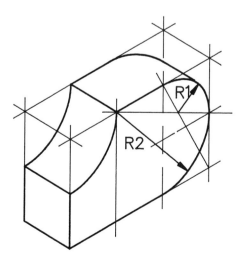

Figure 4-10 Drawing an isometric arc as a portion of an ellipse.

Figure 4-11 The steps in constructing an isometric drawing.

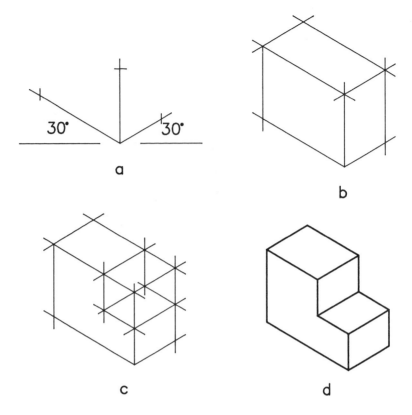

shape of the object, are constructed within the box framework. Last, any missing details are added.

Begin by drawing the three isometric axis lines. You may find it easiest to start with a horizontal line and then lay out the axes at angles of 30° to it (see Figure 4-11a). Next, measure the maximum (or extreme) outside height, width, and depth dimensions; mark them on the three axes. Complete the box with isometric lines as shown in Figure 4-11b. In the next step, construct the object within the box framework, as shown in Figure 4-11c. Complete the drawing by darkening all object lines and erasing construction lines as needed.

Figure 4-12 illustrates the same steps and the same object drawn as it would appear when viewed from the bottom. Though it is permissible, and often useful, to view objects from directions other than the top, it is important to be careful in choosing a viewing direction, or object orientation. Always choose the one that shows the most detailed information about the object. In other words, the view in which the least amount of detail is hidden is the best view. In comparing Figures 4-11 and 4-12, it is obvious that Figure 4-11 gives the most information; therefore, Figure 4-11 is the preferable drawing. Figure 4-13 shows the four orientations possible for an isometric drawing.

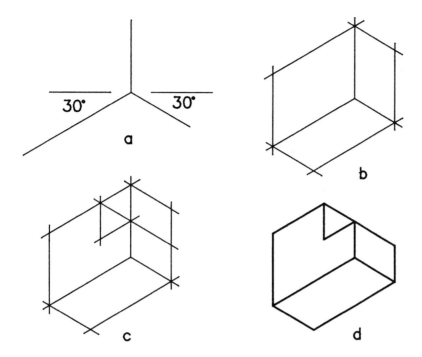

Figure 4-12   The alternate position for an isometric drawing.

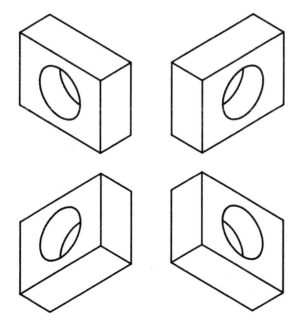

Figure 4-13   Four orientations for an isometric drawing.

The following list summarizes the five basic steps used in constructing an isometric drawing of a simple object. The steps are illustrated in Figure 4-14.

1. Draw the axes that form a front corner of the object.

2. Construct an isometric box whose dimensions match the largest dimensions of the object. Make all measurements along isometric lines.

3. "Carve out" the basic object by measuring and drawing all isometric lines and then constructing all nonisometric lines.

4. Construct any ellipses by using the four-center method.

5. Add centerlines, darken all object lines, and erase construction lines as necessary.

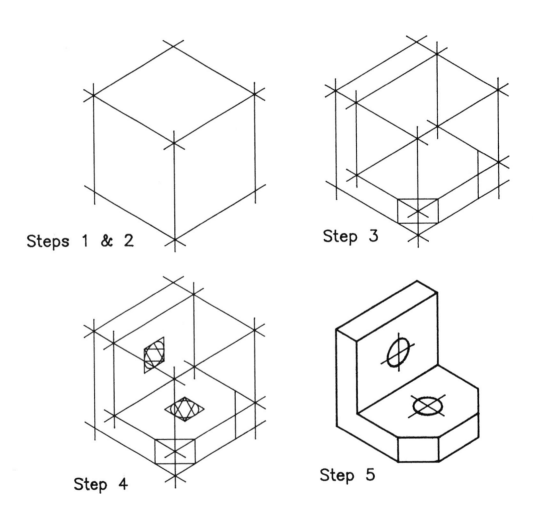

Steps 1 & 2   Step 3

Step 4   Step 5

Figure 4-14 The steps in constructing an isometric drawing.

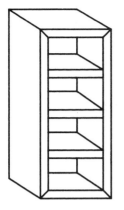 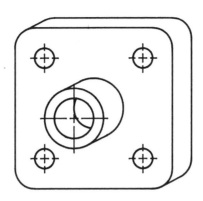

Figure 4-15 Examples of oblique drawings.

4-3 OBLIQUE DRAWINGS

Oblique drawings are constructed so that one face of the object is drawn true size and shape and parallel to the plane of the paper. This makes them quite useful for drawing objects that show most of their detail when viewed from one direction. Figure 4-15 shows several examples of oblique drawings. To give an appearance of depth, **receding lines** are drawn upward or downward, to the right or to the left. All features that appear on the receding faces are distorted in size and shape, but all features that appear on the front face or any surface parallel to it have true sizes and shapes.

The first step in creating an oblique drawing is selecting the most important face (the one that shows the most detail), and drawing it as a flat true-size view. Any surface parallel to the front face is also true size and shape. Select surfaces that have circles and arcs—these will appear as ellipses if drawn on receding faces (see Figure 4-16). Receding lines may be drawn at any angle, but an angle of 30° to the horizontal gives the most realistic appearance. These lines are drawn parallel to each other and at one-half their

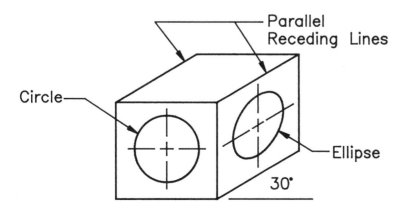

Figure 4-16 An oblique drawing showing the front and receding faces.

Figure 4-17 (a) A cavalier projection, in which the receding lines of an oblique drawing are full-size. (b) A cabinet projection, in which the receding lines of an oblique drawing are half-size.

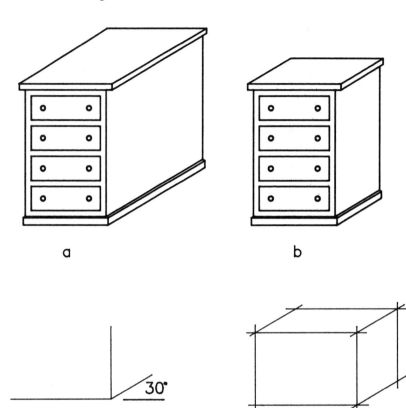

a   b

Figure 4-18 The steps in constructing an oblique drawing.

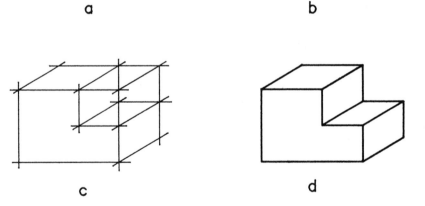

a   b

c   d

true length. If receding lines are drawn full-length, as in Figure 4-17a, the resulting object appears distorted. Oblique drawings with full-size receding lines are called **cavalier projections;** those with half-size lines are called **cabinet projections.**

Like isometrics, oblique drawings are most easily drawn by using the box construction method. Begin by drawing the three axes as shown in Figure 4-18a. Measure and mark the object's outside height, width, and depth on the axes. Complete the box as shown in Figure 4-18b. Add detail by carving out the box as shown in Figure 4-18c. Complete your drawing by darkening lines and erasing as needed. If necessary, ellipses may be constructed by using the

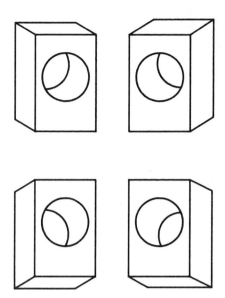

Figure 4-19 Four orientations for an oblique drawing.

four-center method illustrated in Figure 4-7. Remember to use light construction lines until you reach the last step. Like isometrics, oblique drawings may be oriented in several positions, as shown in Figure 4-19.

The following summarizes the five steps used in constructing an oblique drawing. They are illustrated in Figure 4-20.

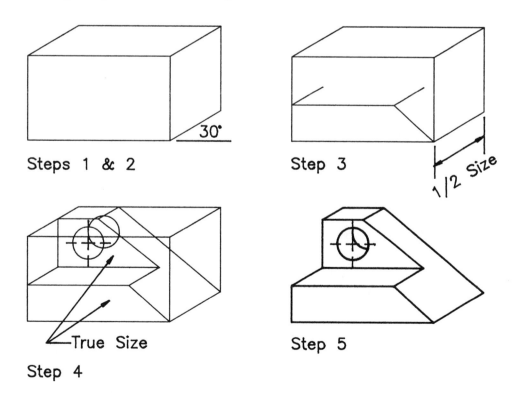

Figure 4-20 The steps in constructing an oblique drawing.

1. Draw the oblique axes that form a front corner of the object.

2. Construct an oblique box whose dimensions match the largest dimension of the object.

3. Draw in the detail of the front face.

4. Project detail along receding lines to appropriate depths and carve out the object.

5. Add centerlines, darken all object lines, and erase construction lines as necessary.

4-4 PERSPECTIVE DRAWINGS

Perspective drawings are the most realistic of the pictorials. Figure 4-21 shows a perspective drawing of a highway. The lanes appear to come together in the distance at a point called the **vanishing point.** This gives the drawing a realistic appearance because objects do look smaller at a distance than they do close up. An isometric or oblique drawing of a highway would look dis-

Figure 4-21 A parallel perspective drawing illustrating a single vanishing point.

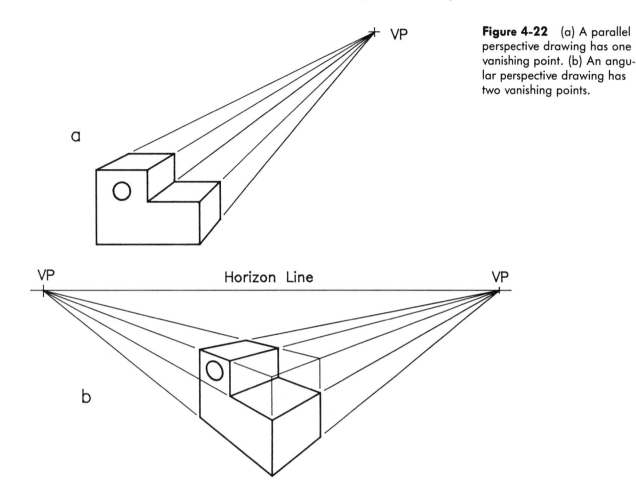

Figure 4-22 (a) A parallel perspective drawing has one vanishing point. (b) An angular perspective drawing has two vanishing points.

torted because the lanes would never converge but would be drawn the same distance apart for their entire length. For this reason, perspectives are better than isometrics or obliques for showing large objects such as buildings and landscapes.

Two commonly used types of perspective drawings are shown in Figure 4-22. These are parallel (one-point) and angular (two-point) perspectives. Notice that the **parallel perspective** has one vanishing point and resembles an oblique drawing. The **angular perspective** has two vanishing points and resembles an isometric drawing, except the receding lines come together instead of remaining parallel.

To draw perspectives of either type, use the procedures for creating isometric and oblique drawings and then establish one or two vanishing points. The vanishing points may be placed anywhere, from above to below the object. However, for angular perspectives, both points must be on a level horizon line as shown in Figure 4-22b. Only the front face of the object in parallel perspective and the front edge in angular perspective can be measured. All receding line measurements must be estimated. Circles and arcs on receding faces do not actually appear as true ellipses, but they are enough like ellipses that ellipses can be used.

4-5 SHADING

Shading can add much to the clarity and readability of pictorial drawings, especially for objects with curved surfaces or complex detail. Unfortunately, effective shading requires time and some artistic expertise. However, a few simple techniques can be used to indicate where internal and external corners are rounded rather than sharp. The shading marks shown in Figure 4-23 are normally done freehand. This type of shading can be used on any type of pictorial drawing.

4-6 CONCLUSION

This chapter has covered the techniques for drawing isometrics, obliques, and perspectives. The drawing of pictorials with instruments has been emphasized. However, the methods presented can easily be adapted to freehand sketching. In your work you may find it helpful to sketch pictorials of the objects you are trying to visualize and to draw. Freehand pictorial sketches are especially useful tools for visualizing and planning multiview drawings.

Check your general understanding of this chapter by answering the following questions.

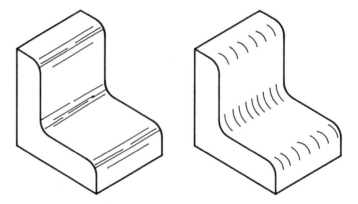

Figure 4-23 Shading techniques for depicting round corners.

REVIEW QUESTIONS

1. The most realistic and most difficult type of pictorial to draw is the:

a. oblique

b. isometric

c. perspective

d. multiview

2. The most widely used pictorial in technical drawing is the:

a. oblique

b. isometric

c. perspective

d. multiview

3. The isometric axes are three intersecting lines with equal angles of _____ between them.

a. 45°

b. 60°

c. 90°

d. 120°

4. Isometric lines are drawn:

a. one-half true length

b. true length

c. one and one-half times true length

d. twice true length

5. Angles in isometric drawing _____ appear true size.

a. always

b. seldom

c. never

d. none of the above

6. Circles and arcs in isometrics appear as:

a. parabolas

b. ellipses

c. cycloids

d. circles and arcs

7. Nonisometric lines are drawn:

a. by using direct linear measurements

b. by using offset linear measurements

c. by using direct angular measurements

d. by guessing

8. Receding lines on oblique drawings are best drawn at an angle of _____ and _____ true length.

    a. 30°; one-half

    b. 30°; their

    c. 45°; one-half

    d. 45°; their

9. All features on the front face of an oblique drawing are shown:

    a. true size and shape

    b. foreshortened

    c. one-half true size

    d. distorted

10. Both isometrics and obliques are best constructed by using the _____ construction method.

    a. geometric

    b. projection

    c. box

    d. four-center

11. The most important step in drawing any pictorial is:

    a. constructing the ellipses

    b. darkening the line work

    c. making measurements

    d. selecting the orientation of the object

12. A principal difference between perspectives and other pictorials is the use of:

    a. box construction

    b. vanishing points

    c. ellipses for circles

    d. relative object orientation

PROBLEMS

4-1 Shown below are four objects drawn as obliques. Draw them as isometrics. Scale the drawings for dimensions and remember that receding lines are drawn one-half size on oblique drawings.

a

b

c

d

4-2 Shown below are four objects drawn as isometrics. Draw them as obliques. Scale the drawings for dimensions and remember that only isometric lines are true size.

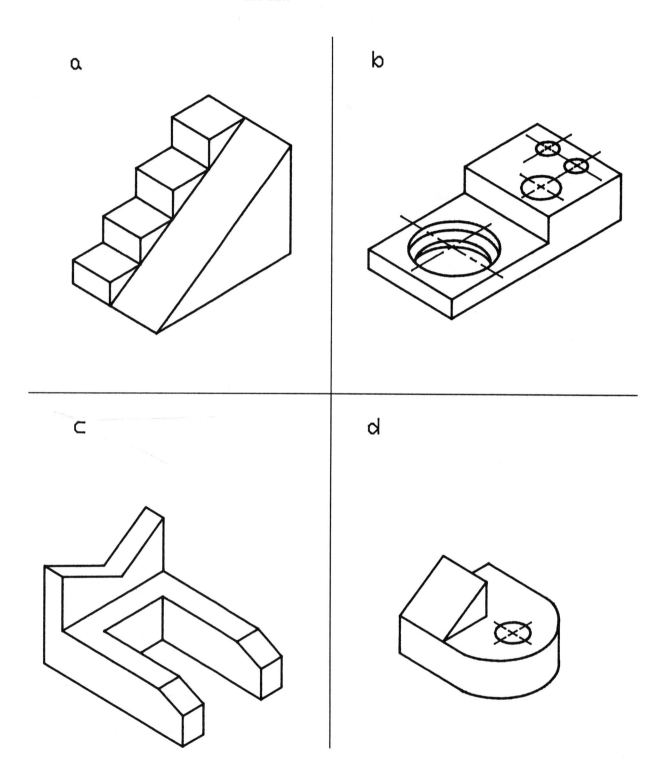

a

b

c

d

4-3 Construct oblique drawings of these objects. Dimensions are in inches.

a

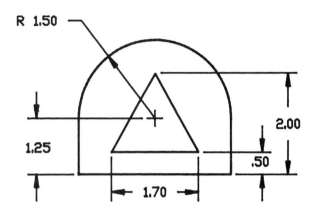

b

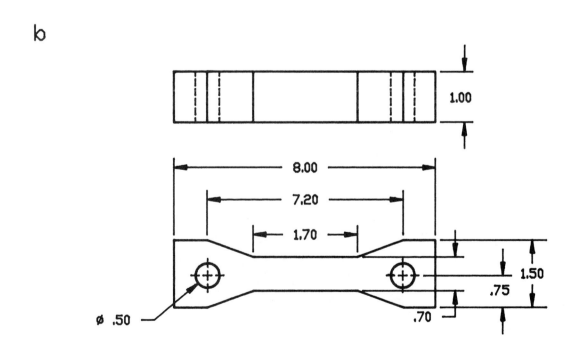

4-4   Construct isometric drawings of these objects. Dimensions are in millimeters.

a

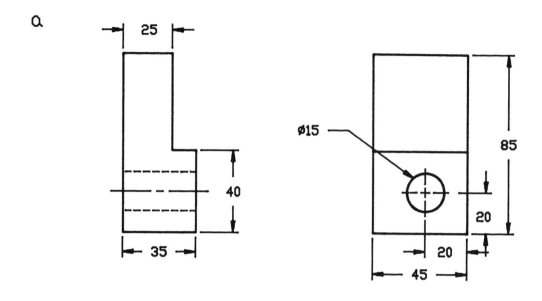

b

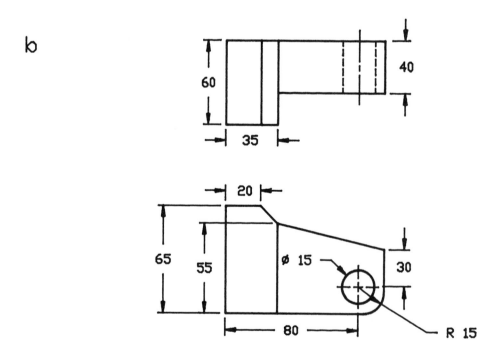

4-5 Construct isometric drawings of these objects. Dimensions are in inches.

a

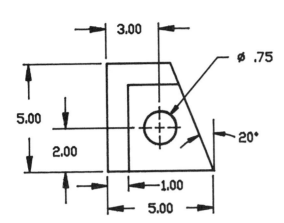

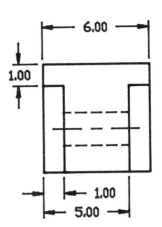

b

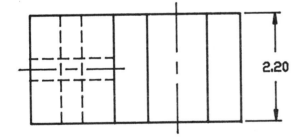

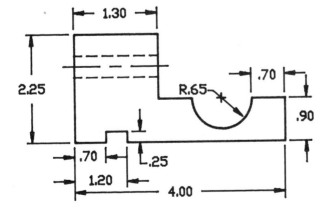

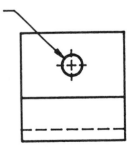

4-6 a. Redraw the objects in Problem 2-12, parts a and b, as obliques. Scale the figures for dimensions.

b. Redraw the objects in Problem 2-12, parts c and d, as isometrics. Scale the figures for dimensions.

c. Redraw the object in Problem 2-13 as an isometric. Follow the dimensions given, but make your drawing one-half scale.

d. Redraw the object in Problem 2-14 as an isometric. Follow the dimensions given, and make your drawing full-size.

e. Redraw the object in Problem 2-15 as an oblique. Scale the figure for dimensions.

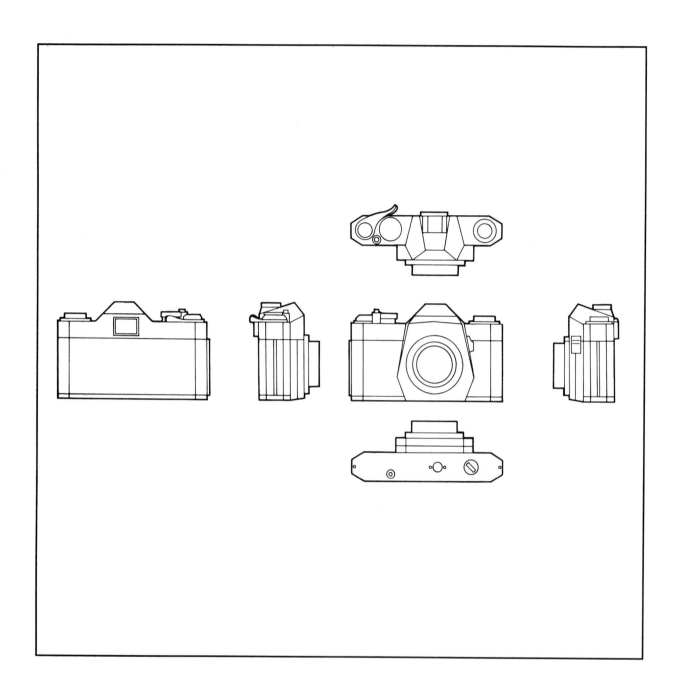

PRINCIPLES OF MULTIVIEW DRAWING

5-1 INTRODUCTION

Have you ever noticed how distorted your view of the world is? Look at the objects around you. Close one eye and estimate their size by sighting on them with your thumb and forefinger. You can see that none of them appears actual size. For example, a person who is across the room looks only an inch tall, but when the same person stands nearer to you, he may appear 5 or 6 in. tall. These optical distortions have probably never bothered you because you, like most people, have learned to mentally compensate for them. Knowledge and experience usually modify what the sense of sight records. However, this mental compensation works only when you already know the size and shape of the object you are viewing. For example, you probably knew that the person across the room was closer to 6 feet tall than 1 in. tall.

In technical drawing the primary purpose is to create graphical descriptions of physical objects whose sizes and shapes are not known. To be usable, these descriptions must be complete and accurate. This means, among other things, that technical drawings must show objects as they really are and not as they appear with all the usual visual distortions. Unfortunately, as Chapter 4 emphasized, pictorial drawings won't do because they contain their own distortions of size and shape.

This chapter examines **multiview drawing,** the principal method used to eliminate the visual distortions and optical illusions found in most other types of drawings. Multiview drawings show objects by means of flat, two-dimensional views, which show the actual size and shape of the objects. However, a three-dimensional object is difficult to describe with a single two-dimensional view. Multiview drawings, as the name indicates, overcome this problem by showing two or more different views of the object. To show com-

After you complete this chapter, you should:

1. understand orthographic projection and the concept of the projection box
2. know the three standard views
3. be able to select a principal view
4. be able to identify a surface on an object as being visible, hidden, or an edge
5. know the three meanings of object lines
6. be able to identify normal, inclined, or oblique surfaces
7. know the precedence of lines

plex objects, it is usually necessary to draw many ("multi") views to achieve a complete and accurate description. Each separate view shows how the object appears from a different direction. The illustration that opens this chapter shows six different views of a camera. In each view you see a different side, but the views are arranged so that you can mentally connect them to form a three-dimensional image of the object. The process of forming this mental image is called visualization, and it is one of the principal focuses of this chapter.

5-2 ORTHOGRAPHIC PROJECTION

In creating a multiview drawing, the drafter imagines that the object is placed in a box with transparent sides, as shown in Figure 5-1. The drafter pictures what can be seen through the sides of the box. These pictures, which are mentally projected onto the sides of the box, are called **views,** and the sides themselves are called **projection planes,** image planes, or picture planes. This entire process is called **orthographic projection,** a term sometimes used in place of *multiview drawing*. "

In orthographic projection you, the observer, are always on the outside of an imaginary projection box looking in, with a line of sight perpendicular to one of the sides so that you see the object through that projection plane. To see different views of the object, you must change your position and look through different projection planes. Since the standard projection box has six sides, six views can be drawn for any object. All views must be shown on

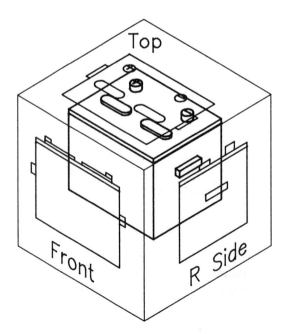

Figure 5-1 A 12-volt battery in an imaginary box. The drafter mentally projects each view onto a projection plane.

one flat sheet of paper, and thus the box is "hinged" so that the sides can be folded out into a single flat plane. Figure 5-2 shows a projection box unfolding, and Figure 5-3 shows the projection box from Figure 5-1 completely unfolded. Carefully study the hinging of the box and the arrangement of the views until you thoroughly understand their relationship to the object and to each other.

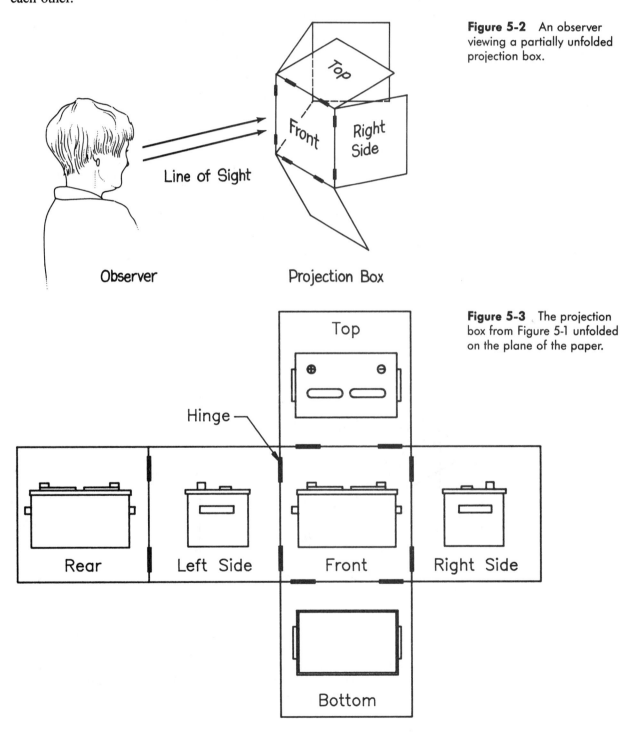

Figure 5-2 An observer viewing a partially unfolded projection box.

Figure 5-3 The projection box from Figure 5-1 unfolded on the plane of the paper.

You probably noticed that the projection planes in Figure 5-3 are labeled. The labels are "Top," "Front," "Right Side," "Left Side," "Rear," and "Bottom." These names refer to the type of view each projection plane represents from the observer's perspective. They are used so that you can identify which view is being referred to in discussions about any object or its drawing.

The placement of an object inside the imaginary projection box is done in a logical manner. Many objects are basically right rectangular prisms, meaning they have square corners, and the sides are rectangular and meet at right angles. When a rectangular prism is placed inside a projection box, the sides can be "lined up" with the sides of the projection box. This places each surface of the prism parallel to one of the projection planes and perpendicular to the line of sight for that view. In Figure 5-4a surfaces A and B are parallel to the top projection plane, surfaces C and D are parallel to the right side projection plane and surfaces E and F are parallel to the front projection plane. The other surfaces will be disregarded for now. However, it may be useful to note they are also parallel to their respective projection planes.

Figure 5-5 shows an object with two slanting surfaces, B and D. It is not possible to orient this object so that all surfaces are parallel to a projection plane. In instances like this the object is oriented so that as many surfaces as possible are parallel to a projection plane. In this way the object can be shown in the least number of views.

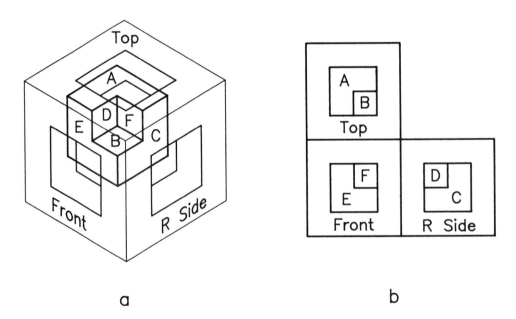

a   b

Figure 5-4 (a) This pictorial drawing shows that surfaces A, B, C, D, E, and F are parallel to their respective projection planes. (b) A multiview drawing that shows relationships among the surfaces.

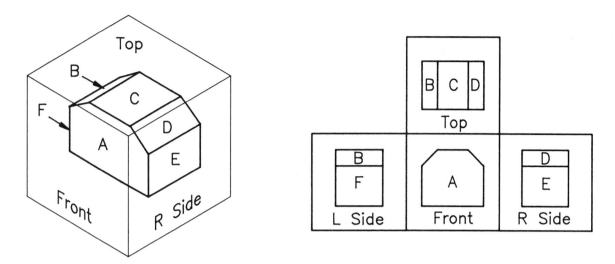

Figure 5-5 Orienting this object so as many surfaces as possible are parallel to a projection plane ensures that the object is shown in the least number of views.

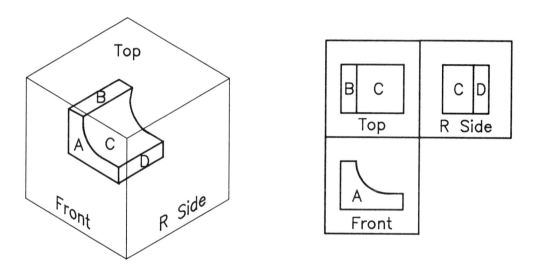

Figure 5-6 An object with a curved surface oriented so the maximum number of surfaces are aligned with projection planes.

Figure 5-6 illustrates the same principle applied to an object having a curved surface, C. The curved surface cannot be parallel to any projection plane. However, the object is oriented so that all the other surfaces are aligned with at least one projection plane.

5-3 THE STANDARD VIEWS

Notice in Figure 5-3 that some of the views, such as the front and rear, look nearly the same. It would be a waste of time and effort to draw both these views, since they give the same information. Identical views of an object may be omitted without sacrificing any clarity. An effective multiview drawing

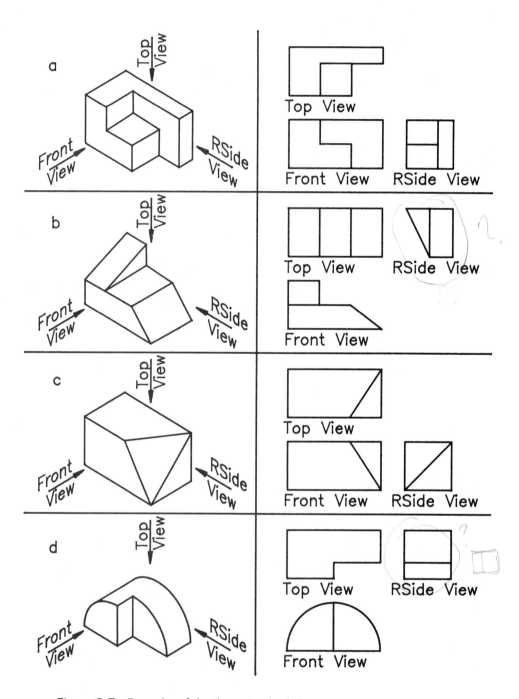

Figure 5-7   Examples of the three standard views.

always contains the minimum number of views necessary to clearly and fully describe the object, but no more. For simple objects the three views normally drawn are the front, top (also called the plan view) and right side (also called the profile). These are referred to as the **three standard views.** Several examples are shown in Figure 5-7.

The bottom, left side, and rear views are often eliminated because they frequently show the same features as the top, front, and right side views. However, it is not always necessary to use even all three of the standard views to completely describe an object. Sometimes, as shown in Figure 5-8, only one or two views are sufficient. On the other hand, extremely complex objects often require many more than three views to completely describe them.

You probably noticed that the orthographic views in Figure 5-7 were shown without the outlines of the projection planes. This is the usual way of drawing multiviews, but some illustrations will continue to show the projection planes to help you understand the examples better.

Look at the multiview drawings for parts b and d in Figure 5-7. Notice that the right side views are in different positions than in parts a and c. This alternate location for the side view is commonly used in industry. It results from hinging the right side projection plane to the top projection plane instead of

a. Single—view drawing of a flat plate with notches & a hole in it. The thickness is shown by a note.

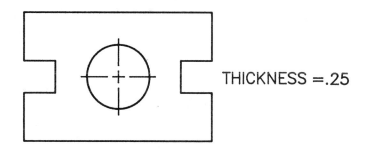

THICKNESS =.25

b. Two—view drawing of a cylinder with notches cut in one end.

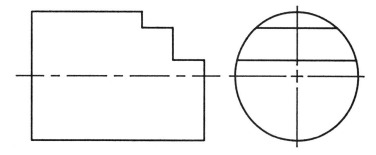

Figure 5-8 (a) A single-view drawing and (b) a two-view drawing. These objects are so simple that all three standard views are unnecessary.

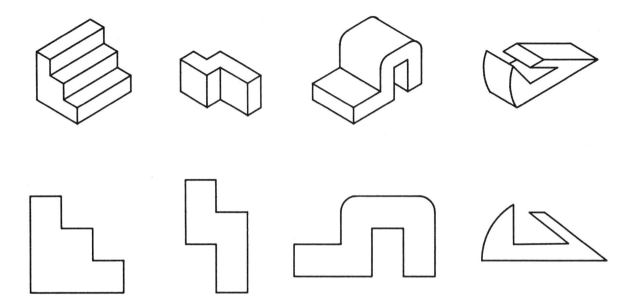

Figure 5-9 The bottom row contains the principal views of the four objects in the top row.

the front projection plane as was shown in Figure 5-2. This second position for the side view will be used frequently throughout this text.

The first and most important view drawn in a multiview is called the principal view. This is the view that most graphically shows the characteristic shape of the object. The principal view is usually drawn as either the top or front view. Figure 5-9 shows several examples.

5-4 READING LINES

All solid objects are composed of, or bounded by, either flat or curved surfaces. How can one surface be distinguished from another on a multiview drawing? In a photograph you can see highlights and shadows, which give clues about an object's shape. On multiview drawings there is no shading; there are only lines to work with. But these lines have specific meanings (see Section 2-6), and if you understand those meanings then you can read multiview drawings.

It is important to note that lines may appear in three different ways. That is, they may appear true length, foreshortened, or as a point. A line is true length (TL) when the projection plane for a particular view is parallel to that line. A line appears as a point when the projection plane is perpendicular to the line. A line neither parallel nor perpendicular to its projection plane is foreshortened. In Figure 5-10 line 1,2 appears as a TL line in the front and

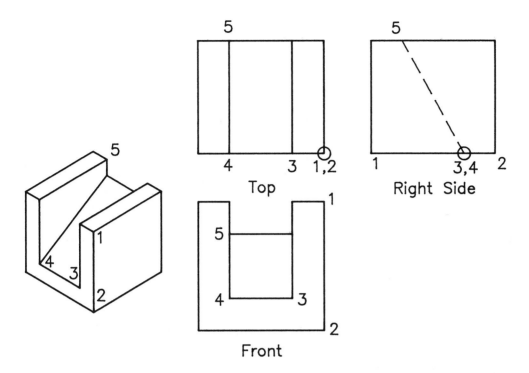

Figure 5-10   TL lines, foreshortened lines, and lines as points.

right side views (note that it is the same length in both views), but as a point in the top view. Line 3,4 appears as a TL line in the front and top views but as a point in the right side view. Line 4,5 appears as a line in all three views, but its length is different in each. Line 4,5 is TL in the right side view because it is parallel to that projection plane; it is foreshortened in the top and front views. Note that the TL view of a line is the longest that it can appear. In any other view the line appears shorter than it actually is.

Object Lines

Lines on technical drawings that show the size and shape of the object are called **object lines.** Object lines can represent three kinds of conditions or features:

- the edge view of a surface
- the line of intersection between two surfaces
- the limiting element of a curved surface

The object shown in Figure 5-11 shows object lines being used in all three ways. Surface A, which is seen as a surface in the top view, appears as an edge (line 1,13) in the front view and as line 14,15 in the right side view.

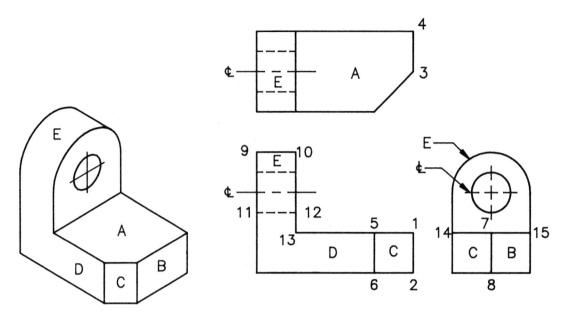

Figure 5-11   Examples of object lines and centerlines.

Surface B is seen as line 1,4 in the front view and line 3,4 in the top view. These examples illustrate cases in which a surface, seen as an edge, is represented by an object line. Line 5,6 in the front view does not represent the edge view of any surface—it has a different meaning. Line 5,6 is the line of intersection between surfaces C and D. Likewise, line 7,8 in the right side view is the line of intersection between surfaces B and C. These examples illustrate cases in which an object line represents the intersection between two surfaces. Line 9,10, seen in the front view, is a case in which an object line shows the limit of a curved surface. Looking at the right side view, you can see that surface E (seen as an edge) is curved rather than flat. Line 9,10 in the front view represents the position where the curve reaches its peak, or limit.

Line 11,12, in the front view, has the same meaning as line 9,10; line 11,12 is also the limit of a curved surface, which, in this instance, is the inside of a round hole. But this surface is actually hidden from the line of sight, and so it is represented by dashed lines (see Section 2-6). These are called hidden object lines. They have the same meaning as visible object lines, except they are used wherever an edge view of a surface, an intersection of two surfaces, or a limit of a curved surface occurs on either the inside or the far side of an object.

Centerlines

Refer again to Figure 5-11. As you know, the lines drawn with alternating long and short dashes and indicated by " ¢ " in each view are called centerlines (see

Section 2-6), and they are used to show symmetry, or the center location of an object or feature. In this example they indicate that the hole is round and where its center is.

The Precedence of Lines Principle

The precedence of lines principle says that, if two or more lines overlap one another exactly, only one line appears in that location. Line 1,2, seen in the right side view of Figure 5-12, is a visible object line representing the edge view of surface A, which lines up exactly with the lower limit of the hole. In the right side view, only the visible object line appears. In the left side view, only one hidden line appears. The visible object line representing surface A and the hidden object line representing the limit of the hole are said to be coincident. Whenever this occurs, only one line can be shown and there is a specific order of precedence. Visible object lines always take precedence over hidden object lines, and hidden object lines take precedence over center-lines. Notice also that line 3,4, which is the line of intersection between surfaces B and C, is coincident with the upper limit of the hole. According to the precedence of lines principle, line 3,4 takes precedence in the left side view and only one hidden line appears in the right side view. A visible object line takes precedence over a centerline, as line 5,6 in the top view shows.

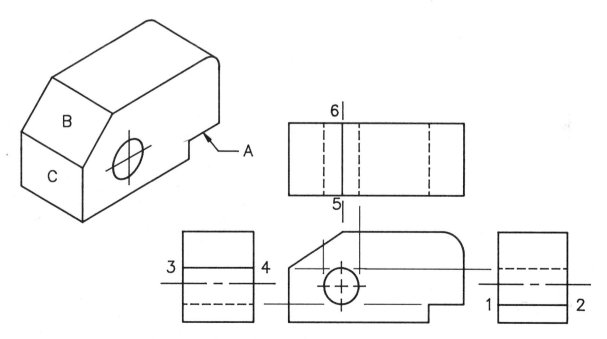

Figure 5-12 Examples of the precedence of lines.

5-5 READING SURFACES

You recall that solid objects have three dimensions and that each view in a multiview drawing shows only two of those dimensions. It might surprise you to know, then, that each view shows every surface of the object. Some show as visible surfaces, some as hidden surfaces, and some as edges. By looking at only one view, you can see all the surfaces that make up the object, but you cannot determine their exact size or shape until you look at the other views and visualize the object from the information given in at least two of them. For example, surfaces may be either flat or curved, but you cannot know which until you see the surface as an edge.

Study Figure 5-13 carefully and try to identify each surface in each view. Notice that surface A appears as a visible surface in the top view and as an edge (line) in the front and side views. Referring to the projection box, you can see that surface A is parallel to the top projection plane and perpendicular to the front and side projection planes. Surface A is called a **normal surface,** and it appears true size and shape in the top view. **True size and shape** (abbreviated **TS&S**) means that all lines are true length and all angles are true size. Note that a surface parallel to a projection plane is also perpendicular to the lines of sight for that view. Given three standard views, a normal surface always appears TS&S in at least one view (not necessarily the top) and

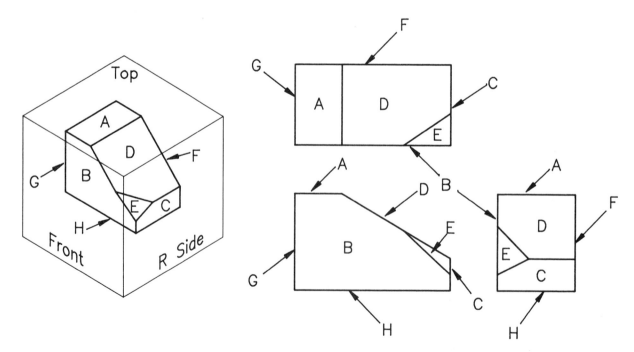

Figure 5-13   An object with normal, inclined, and oblique surfaces.

as an edge in the other two. What other surfaces are normal in Figure 5-13? How about B and C? B shows TS&S in the front and C shows TS&S in the side view. How about surfaces F, G, and H? All are normal surfaces. The difficulty in recognizing them is that they appear as hidden TS&S surfaces. For example, in the front view, surface F is TS&S, but it is hidden behind surfaces B and E.

You have seen that, in relation to a projection plane, normal surfaces are either parallel, in which case they appear TS&S, or perpendicular, in which case they appear as an edge. Looking at the projection box in Figure 5-13, you can see that surface D is perpendicular to the front projection plane but not parallel or perpendicular to the top and side projection planes. In the multiview drawing you can see that surface D shows up as an edge in the front view (remember, it was perpendicular to the front projection plane) and as a foreshortened surface in the top and side views. Notice that the shape of D is similar in both views, but the apparent size is quite different. This type of surface is called an **inclined surface.** To see it TS&S requires a special type of view called an auxiliary view. (Auxiliary views are covered in Chapter 7.) In any standard three-view drawing, an inclined surface shows once as an edge and twice as a foreshortened surface.

Surface E in Figure 5-13 is the third surface type, the **oblique surface.** In the projection box, you can see that it is neither parallel nor perpendicular to any of the projection planes. In the multiview drawing, surface E never appears as an edge or TS&S. It appears foreshortened in all three views. It appears a different size in all three views but never true size. Oblique surfaces require two auxiliary views to find them TS&S.

However, notice that surface E retains its triangular shape in all three views. This illustrates an important principle: All surfaces retain their basic shapes in all views, except when they appear as edges.

5-6 VISUALIZATION TECHNIQUES

Refer to Figure 5-14, which shows a top view of an object. How much information does this one view give you? It tells you the width and depth (but not the height), the general shape (rectangular), and that the top has three different surfaces. You know this because of object lines 3,5 and 4,7; if the top were a single surface, those lines wouldn't be there. (Remember the three kinds of features that object lines can represent.) But from the one view shown in Figure 5-14, you can have no idea of the nature of those three different surfaces. It depends on the meaning of the object lines shown.

Figure 5-14 A top view of an object.

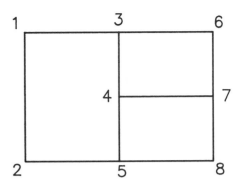

If the lines in Figure 5-14 represent edges of surfaces, the object might look like any one of the examples shown in Figure 5-15. Study the multiviews carefully. Can you visualize the object in each case? Can you see how the object lines in each view relate to the surfaces in the pictorials?

If some of the lines in Figure 5-14 represent intersections of surfaces, then the object could look like any one of those shown in Figure 5-16. Notice that the surfaces all look the same in the top views as they did in Figure 5-15.

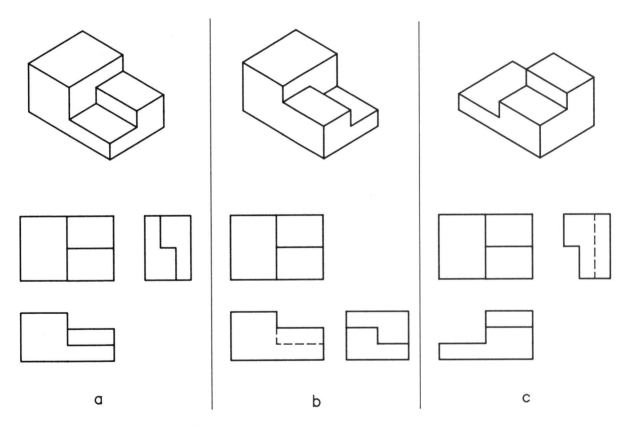

a   b   c

Figure 5-15 Several possible interpretations of Figure 5-14 if object lines represent edges.

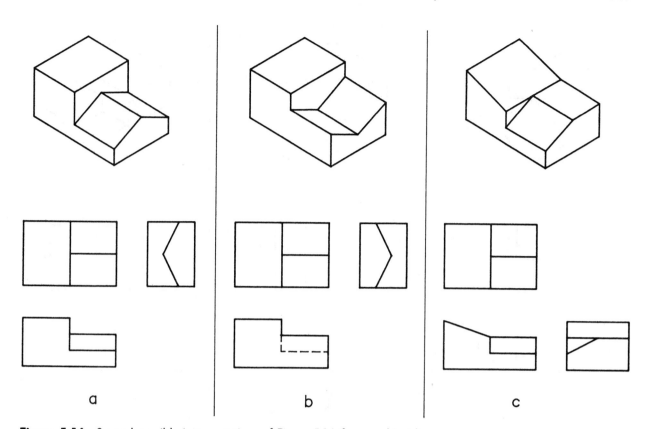

Figure 5-16 Several possible interpretations of Figure 5-14 if some object lines represent intersections.

If some of the surfaces shown in Figure 5-14 are curved, then the objects could look like those seen in Figure 5-17. Notice that you can see that a surface is curved only by studying the edge, or contour view.

The point is, you can't know from one view exactly what an object looks like. However, with two or more views and a knowledge of the meanings of lines, you can see the object in your mind's eye and determine its size and shape. This process is called *visualization*.

Complex objects are often difficult to visualize, even when there are three or more views to look at. This is especially true when there is no pictorial view to refer to, which is normally the case with technical drawings used in industry. However, by using the knowledge gained so far from this chapter, you should be able to visualize most objects. To review:

1. Object lines may be either visible or hidden, and they can have three possible meanings.

2. Lines may appear true length, foreshortened, or as points.

3. Surfaces may be plane (flat) or curved.

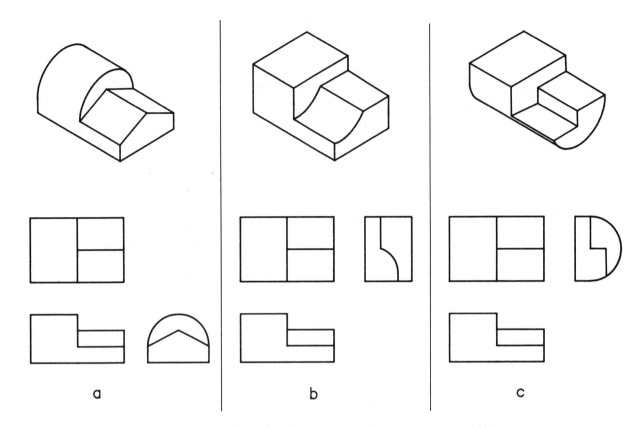

a   b   c

Figure 5-17 Several possible interpretations of Figure 5-14 if some object lines represent curves.

4. Plane surfaces may appear true size, foreshortened, or as edges. Plane surfaces are classified as either normal, inclined, or oblique.

5. Every surface of an object appears in every view as either a visible surface, a hidden surface, or an edge (line).

6. A surface retains its basic shape in every view, even though its apparent size may change (unless it appears as an edge.)

For particularly tough visualization problems, the technique of numbering corners, as shown in Figure 5-18, is often helpful. Looking at the front view, notice that each corner has two numbers. That is because each of these corners, or points, represents the point view of a line and, of course, a line has two ends. In Figure 5-18 the first number belongs to the nearest end point and the second number to the end point farther away. Thus line 1,6 appears as a point in the front view, as the visible line closest to you in the top view, and as the hidden line farthest from you in the right side view. Line 2,7 is also a point in the front view and a visible line in the top view, but is visible rather

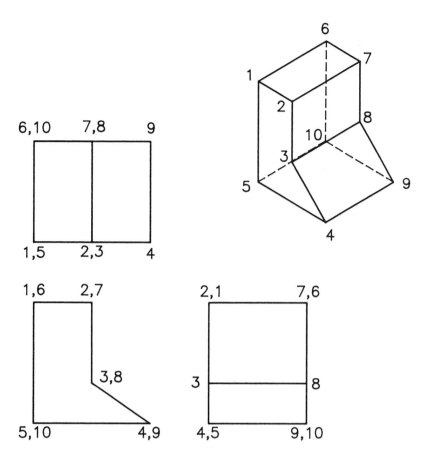

Figure 5-18 Locating and identifying lines and surfaces by their end and corner points.

than hidden in the right side view. Together with line 1,6, it describes the normal surface, 1,2,6,7. Line 5,10 is a point in the front view and a hidden line in both top and right side views. All the lines appear as points, visible lines, or hidden lines.

Inclined or oblique surfaces can also be described by using corner points. Locate surface 3,4,8,9 in Figure 5-18. It is a rectangle, and it appears as such in the top and right side views. Notice that its size is a little smaller in the right side view than in the top view. Since it appears as an edge in the front view, you can identify it as a plane inclined surface. Study Figure 5-18 and identify all the lines and surfaces in each view. If you have any difficulty, use the isometric view to help your visualization.

Look at Figure 5-19. It is a more complex object than the one shown in Figure 5-18, and there is no pictorial. Can you visualize its shape? Check the numbers. Look for point views of lines and then find those lines in the other views. Are they visible or hidden? Look at the surfaces and determine where they show as edges and where they show as surfaces, both visible and hidden. Look for any inclined or oblique surfaces.

Figure 5-19 Numbered points in a multiview drawing of a relatively complex object.

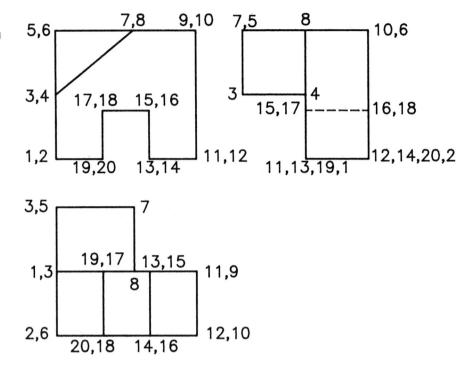

Another method that is useful for visualizing an object from a multiview drawing consists of constructing an isometric box and, in stages, establishing lines and surfaces until you have a pictorial view that explains the multiview. In Figure 5-20 this method is used to establish the shape of the object drawn in Figure 5-19.

5-7 CONCLUSION

This chapter has introduced the basic principles of orthographic projection and multiview drawing. The three standard views and the principal view have been emphasized. The chapter discussed the meanings of object lines and the precedence of lines and introduced the point, true length, and foreshortened views of lines. Plane and curved surfaces; normal, inclined, and oblique surfaces; and true size and foreshortened and edge views of surfaces have also been covered. The emphasis has been on reading multiview drawings. Figure 5-21 contains drawings that illustrate all the concepts contained in this chapter.

Check your general understanding of this chapter by answering the following questions.

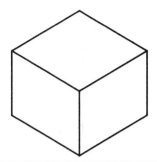

Step 1. Draw an isometric box to the outside dimensions of the object.

Step 2. Draw all indicated lines on the sides of the box corresponding to the orthographic views.

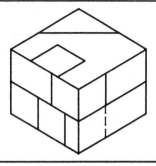

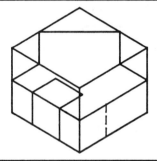

Step 3. Establish actual surfaces and project lines and points onto them.

Step 4. Complete the drawing by adding needed lines and erasing un- needed ones.

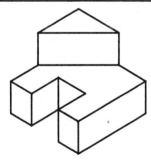

Figure 5-20 The steps in visualizing a multiview by drawing a pictorial.

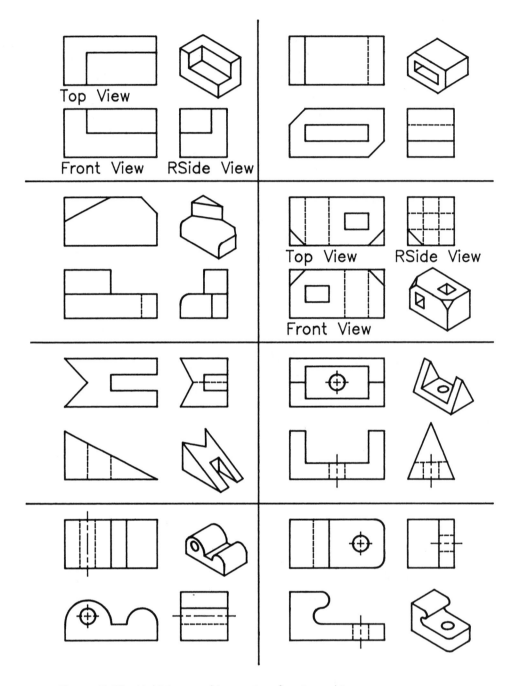

Figure 5-21 Multiviews and isometrics of various objects.

REVIEW QUESTIONS

1. In orthographic projection, lines of sight are always _____ to the projection plane.
   a. perpendicular
   b. parallel
   c. inclined
   d. oblique

2. An effective multiview drawing contains:
   a. three views
   b. six views
   c. a minimum number of views
   d. as many views as possible

3. The first and most important view drawn in a multiview is the:
   a. top view
   b. front view
   c. principal view
   d. main view

4. Which of the following is *not* a possible meaning for an object line?
   a. edge view of a surface
   b. symmetry of a surface
   c. limit of a curved surface
   d. intersection of two surfaces

5. Which of the following is the correct order of line precedence?
   a. center, hidden, visible
   b. visible, hidden, center
   c. hidden, center, visible
   d. visible, center, hidden

6. Centerlines are used to indicate:
   a. none of the below
   b. eccentricity
   c. perpendicularity
   d. parallelism

7. A plane surface perpendicular to one projection plane and *not* parallel to any is called:

   a. a normal surface

   b. an inclined surface

   c. an oblique surface

   d. a bounded surface

8. A plane surface neither perpendicular nor parallel to any projection planes is called:

   a. a normal surface

   b. an inclined surface

   c. an oblique surface

   d. a bounded surface

9. Lines and surfaces appear true size in a multiview only when the line of sight is _____ to the line or surface.

   a. parallel

   b. inclined

   c. oblique

   d. perpendicular

10. Lines and surfaces appear true size only when a projection plane is _____ to the line or surface.

    a. parallel

    b. inclined

    c. oblique

    d. perpendicular

11. Surfaces on an object may appear as:

    a. hidden surfaces

    b. visible surfaces

    c. edges

    d. all the above

12. The least possible number of views that can be drawn of any particular object in a multiview drawing is:

    a. one

    b. two

    c. three

    d. six

PROBLEMS

5-1 Identify each lettered surface shown on the pictorial drawing that follows by the corner numbers shown in the multiview drawing. Enter the corner numbers in a table like the one that follows.

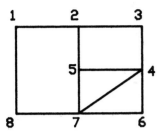

Top

Right Side

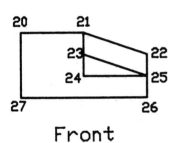

Front

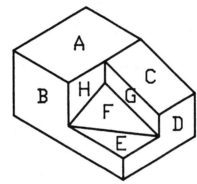

Surface	Top	Front	Right Side
A	1,2,7,8	20,21	10,18
B			
C			
D			
E			
F			
G			
H			

5-2 Identify the lines and planes as indicated in the tables that follow.

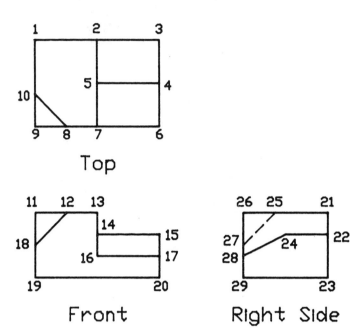

Top

Front   Right Side

LOCATING LINES

Find This Line	In This View	Answers
1–2	F	11–13
1–2	RS	
11–12	T	
4–5	RS	
1	RS	
28–29	T	
18–19	RS	
21	T	
24	F	
26–28	F	
2–3	RS	

LOCATING PLANES

Find This Plane	In This View	Answers
1-2-7-8-10	RS	21-25-26
1-2-7-8-10	F	
14-15-16-17	T	
14-15-16-17	RS	
22-24	T	
25-26-27	F	
14-15	RS	
19-20	T	
21-22-23	F	
15-17-20	RS	
1-9-10	RS	

5-3   Identify each visible surface in each view of the multiview drawing that follows by matching the letters with the corresponding number in the pictorial.
Identify each surface as normal, inclined, or oblique.

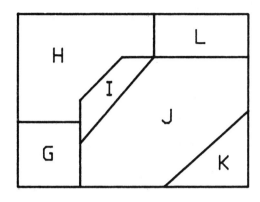

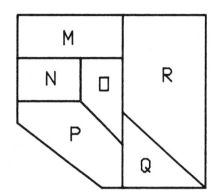

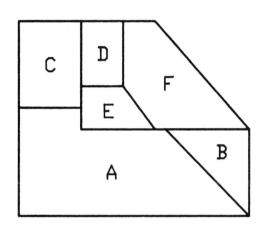

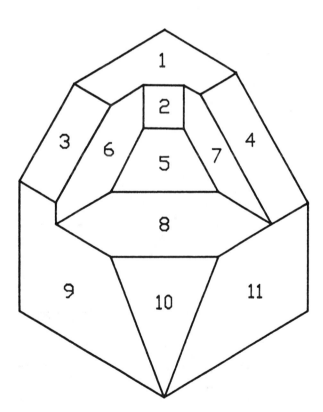

5-4   In parts a and b, identify the meanings of the lines indicated by numerals. Use *I* for intersection of surfaces, *E* for edge view of a surface, and *L* for limit of a curved surface.

a

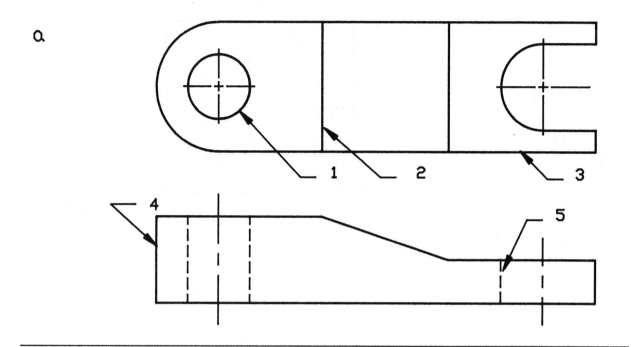

b

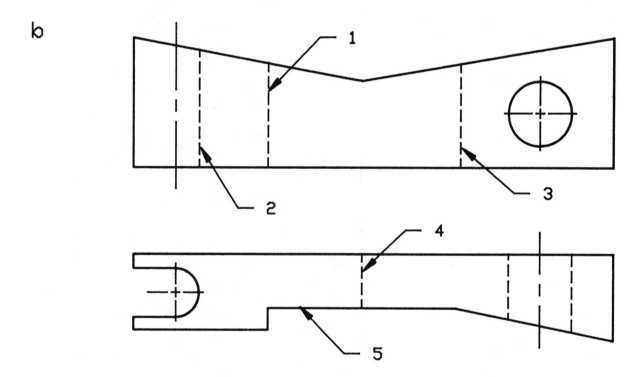

5-5 a. Match the numbered surfaces shown in the top views with the lettered edge
views shown in the front views.

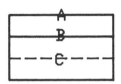
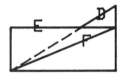
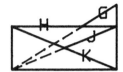
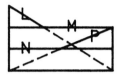

b. From the three views that follow, determine the depth of each hole, A
through F.

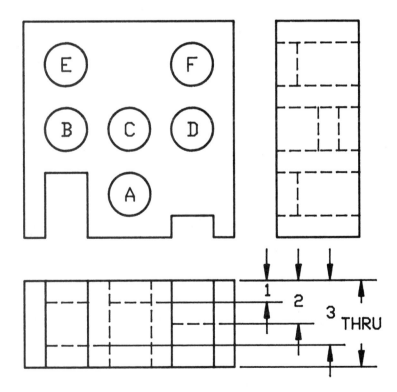

5-6   Using the numbers given in the top view, number each corner on surfaces A, B, C, and D in the front and right side views. Redraw the views on a separate paper.

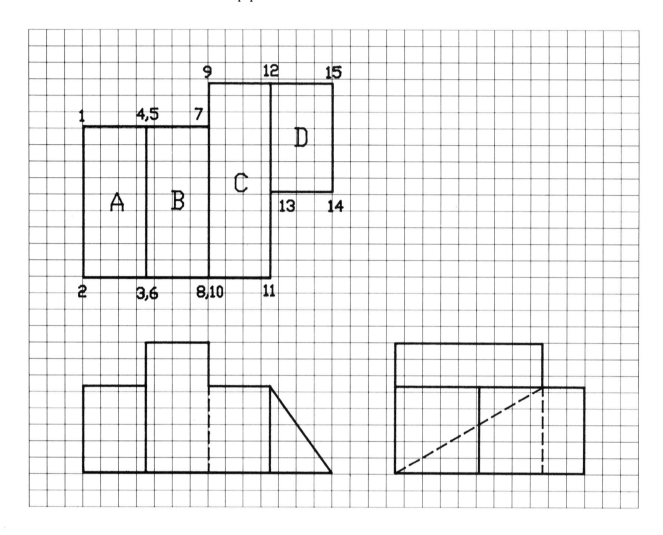

5-7 Redraw the multiviews in parts a and b on a separate paper, completing the front and side views by using the correct types of lines, as indicated by the precedence of lines principle. All holes are through holes.

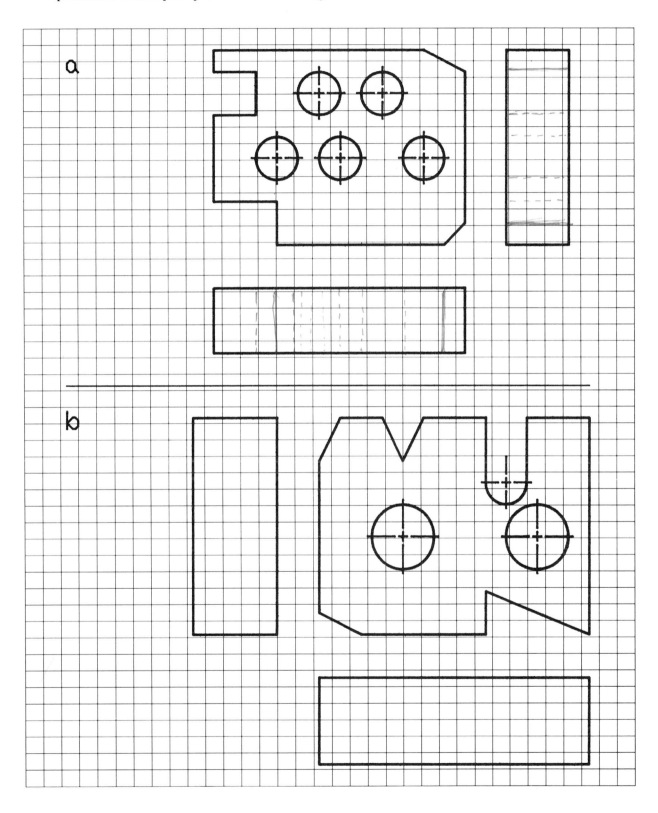

5-8 On a separate piece of paper, match the number of the multiview on the right with the letter of the pictorial on the left that illustrates it.

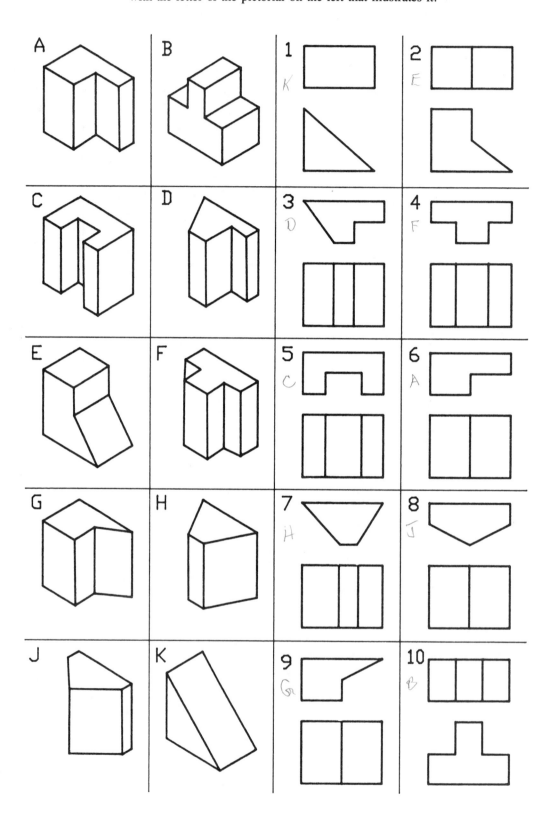

5-9 On a separate piece of paper, match the number of the multiview on the right with the letter of the pictorial on the left that illustrates it.

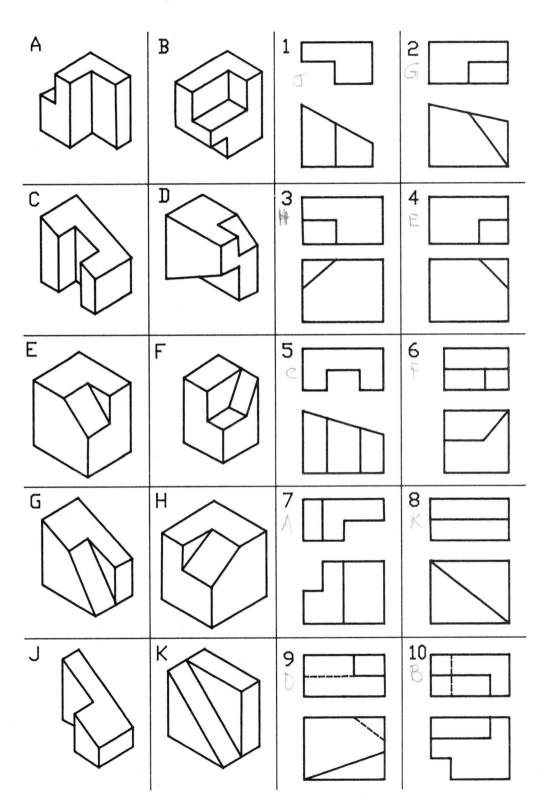

5-10   For each object in parts a and b, select the correct top, front, right side, left side, and rear views from those shown on the right.

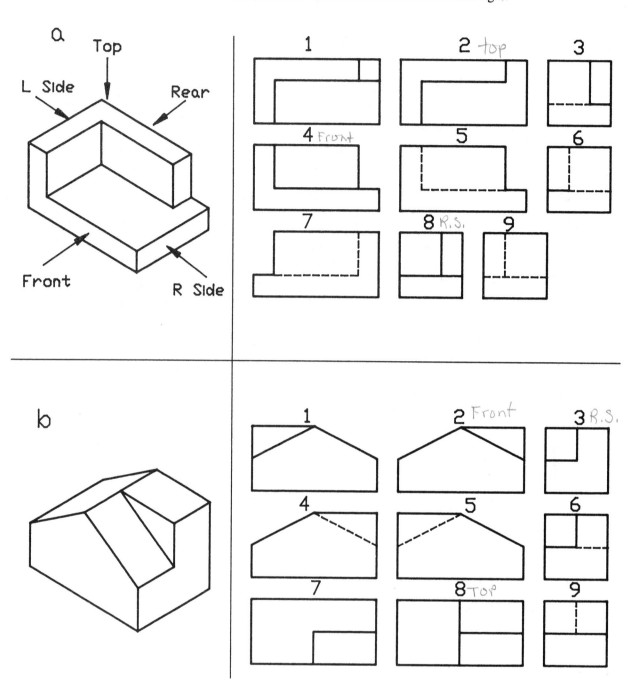

5-11   In parts a through c, select the correct right side view for the two given views.

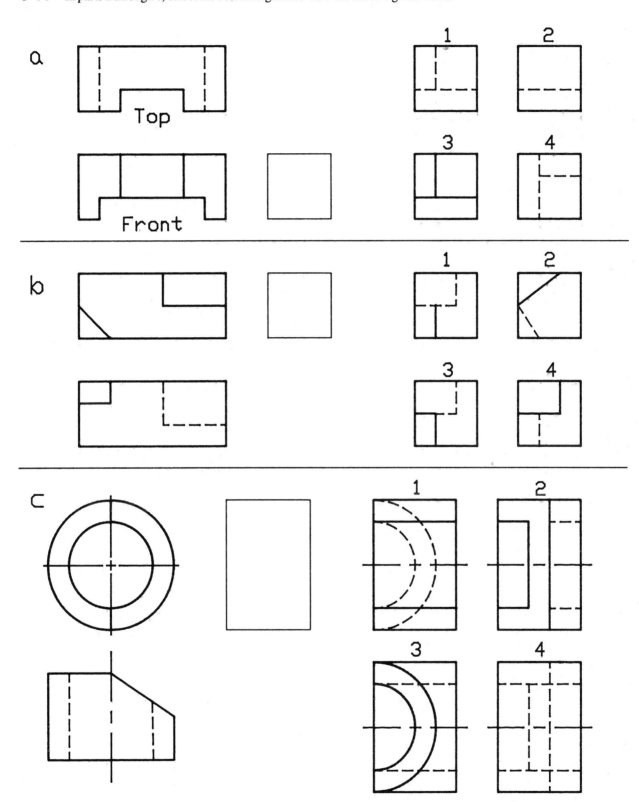

5-12   Select the views that match the views indicated by the arrows.

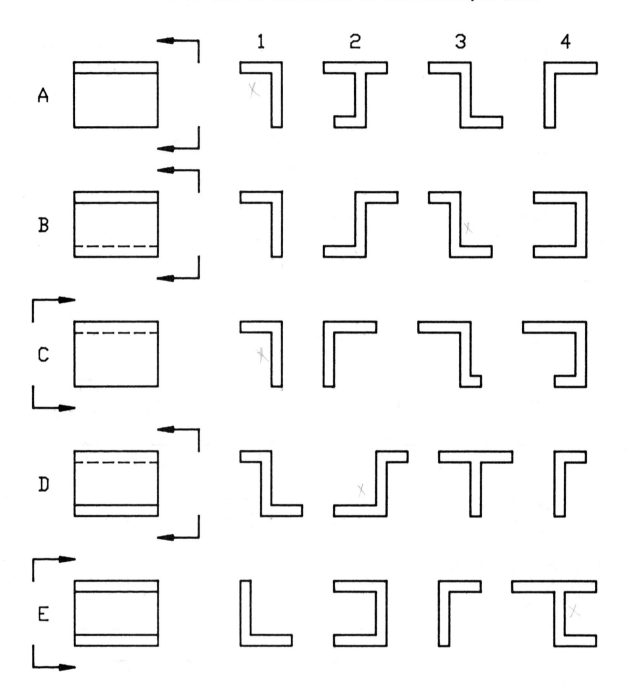

5-13   On a separate paper redraw the multiviews shown in each part that follows. Complete them as necessary by drawing the missing object lines. Some of the views are complete.

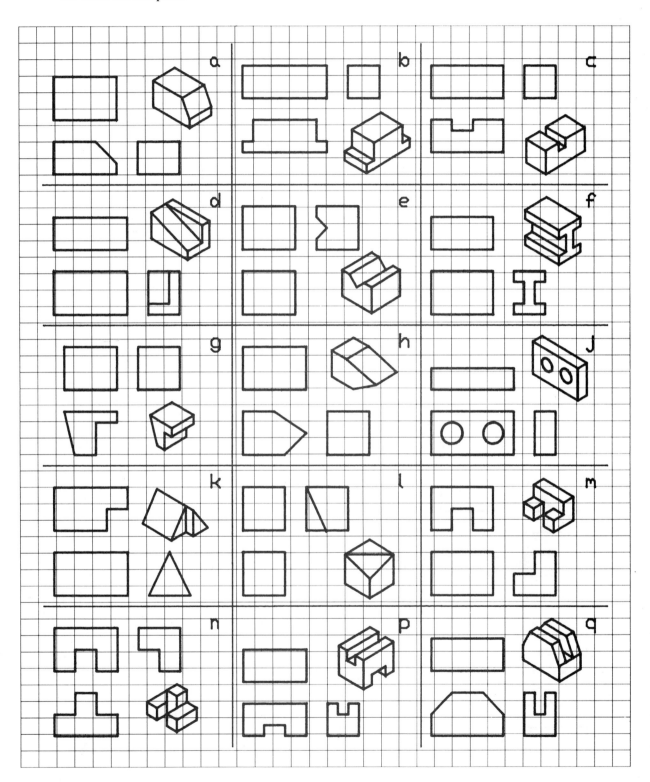

5-14   On a separate paper redraw the multiviews shown in each part that follows. Complete them as necessary by drawing the missing object lines. The labeled views are complete.

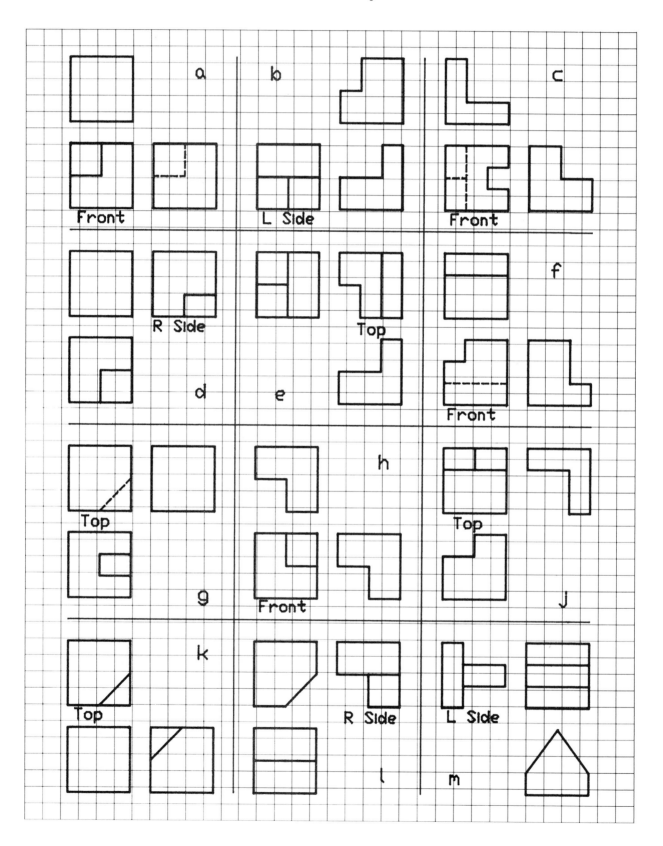

5-15 On a separate paper redraw the multiviews shown in each part that follows. Complete them as necessary by drawing the missing object lines. The top view is complete in each case.

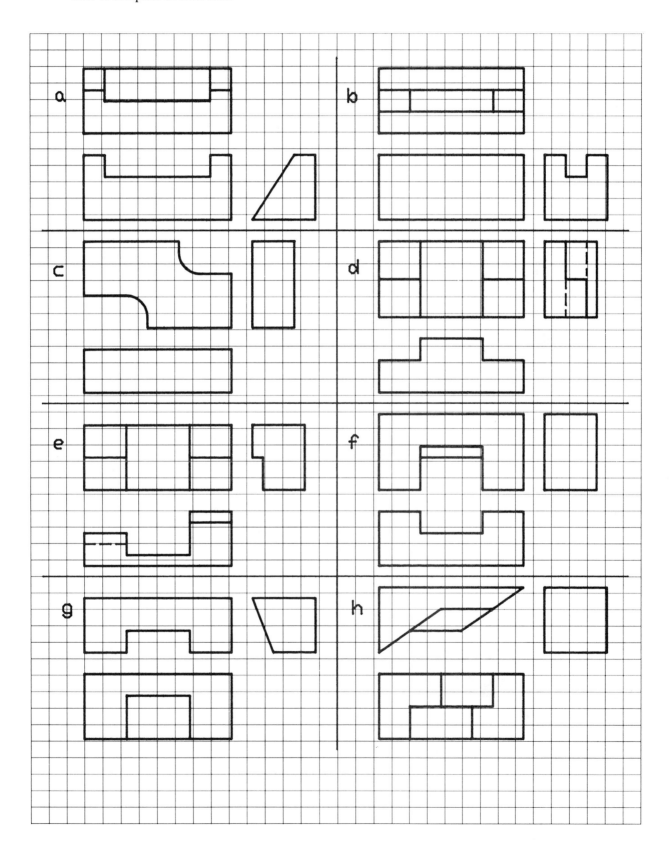

5-16 On a separate paper redraw the multiviews shown in each part that follows. Complete them as necessary by drawing the missing object lines. At least one view is complete in each case.

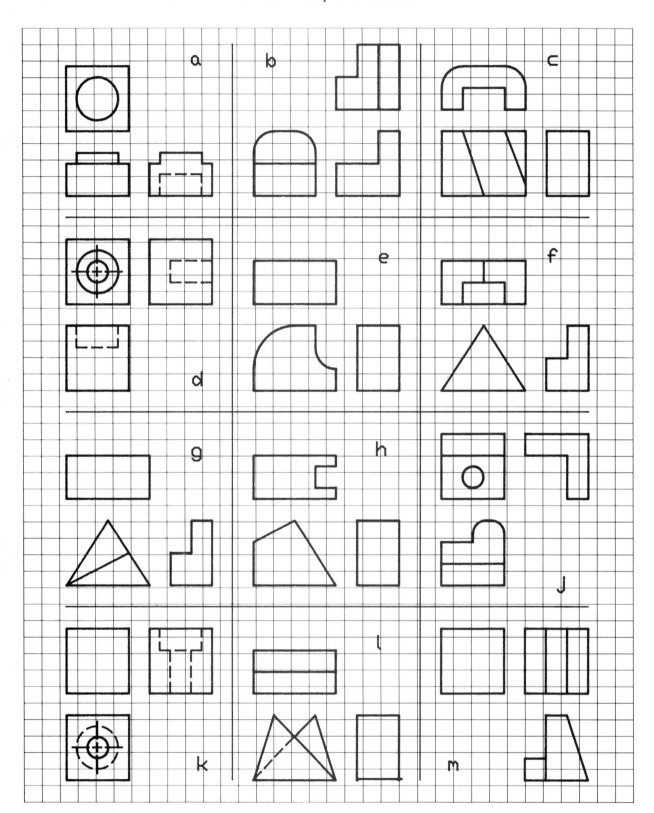

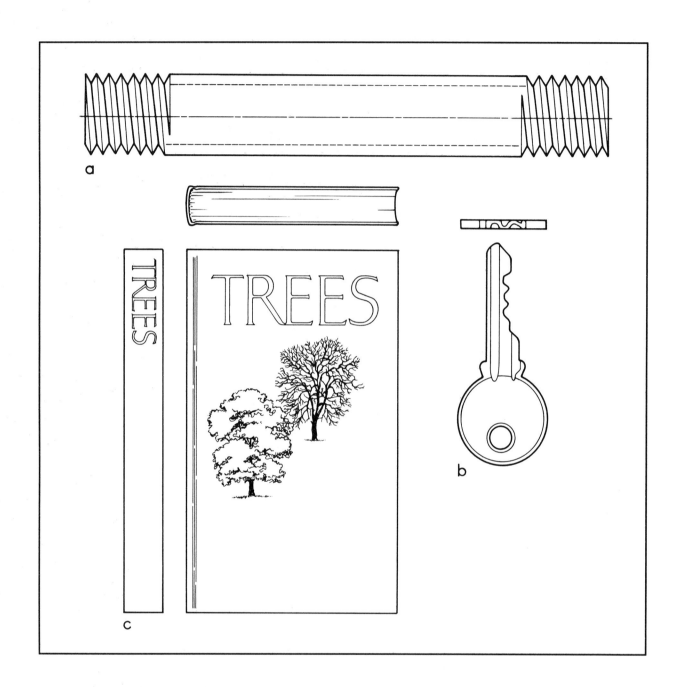

a

b

c

TREES

TREES

BASIC MULTIVIEW DRAWING TECHNIQUES

6-1 INTRODUCTION

The illustration that opens this chapter shows three examples of multiview drawings of common objects that require different numbers of views to adequately describe them. Part a shows a piece of common water pipe. Since you know that it is round, only one view is needed because the diameter as well as the length can be measured in that one view. The second example, b, shows a two-view drawing of a key. The second view is needed to show the thickness and the depth of the grooves. Part c is a three-view drawing of a book. Three views are necessary to clearly describe the shape and appearance (including the lettering) of the book. Each of these examples uses the most descriptive views, the simplest views, and the least number of views that adequately describe the object. How did the drafter know which views to draw?

The last chapter covered the principles of multiviews and how to read them. A thorough understanding of those principles is needed to use technical drawings. But perhaps you want to know how to draw them, also. This chapter will show you how. It will present techniques for constructing one-, two-, and three-view drawings and partial views, and it will introduce the concept of conventional representations.

Simple objects require single-view drawings only, but more complex objects require additional views. Partial views are used to eliminate unneeded details. As the name indicates, partial views show only part of what a standard view would show. Conventional representations include various standard ways of showing features that frequently appear in technical drawings. Of particular importance are the techniques used in drawing hidden lines.

The emphasis in this chapter will be on applications of the most efficient means for drafting readable and accurate drawings.

After you complete this chapter, you should:

1. understand the use of reduced and enlarged scales in drawings

2. be able to construct one-, two-, and three-view multiview drawings

3. understand the relationship between adjacent and related views

4. be able to relate the size dimensions of an object to the number of views required for a clear description

5. be able to draw partial views and know when they should be used

6. know the drawing conventions regarding hidden lines, rounded intersections, and runouts

175

6-2 DRAWING SCALES

With a few exceptions, multiview drawings are always made to some **scale.** This means they are drawn to a certain size relative to the actual size of the object. The scale used depends on the size of the drawing paper and on the size and complexity of the object.

If an object that measures 50 mm × 100 mm is to be drawn, it can easily be drawn at its actual size, or full-size, on any size drawing sheet. But if the object is 300 mm × 500 mm, then it can be drawn full-size on an A2-, C-size, or larger sheet only. (See Table 2-1 for an explanation of sheet sizes.) If it is to be drawn on an A4-, or A-size sheet, it will have to be reduced to at least half its actual size, which would make it 150 mm × 250 mm. This process is called drawing to a reduced scale. This particular drawing would have a scale of 1/2, also written as "half size," "1:2," or 1 mm = 2 mm."

Going the other way, it is often necessary to enlarge the scale of an object to show more detail and thereby gain greater clarity. For example, the 50 mm × 100 mm object might better be drawn twice as large. It would then be drawn 100 mm × 200 mm and the drawing scale would be twice size, 2:1, or 2 mm = 1 mm.

If you think about it, you will quickly realize that many objects must always be drawn at some reduced scale (automobiles, houses, 747s, and the like) and others must be drawn at an enlarged scale (watch gears, integrated circuit chips, sewing needles, and other small items). Table 6-1 shows some commonly used scales and how they are specified on drawings.

Note: Architectural drawings use a completely different set of equivalencies to reduce scales. Fractions and multiples of inches are made equivalent to feet and inches. All architectural drawings are made at reduced scales.

Here is an important thing to remember: Don't try to determine the actual size of an object from the size that it appears in a drawing. Size information is given only by dimensions, which are covered in Chapter 8.

Table 6-1 Commonly Used Drawing Scales

Size	Ratio	Fraction	Equivalency
Full	1:1	1/1	1 = 1
Half	1:2	1/2	1 = 2
Quarter	1:4	1/4	1 = 4
Tenth	1:10	1/10	1 = 10
Twice	2:1	2×	2 = 1
Ten times	10:1	10×	10 = 1

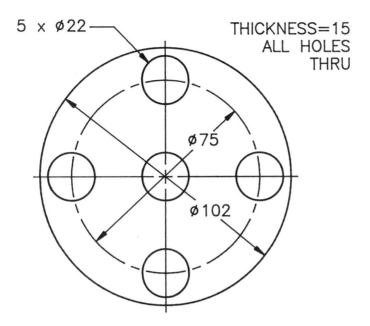

5 × ⌀22

THICKNESS=15
ALL HOLES
THRU

⌀75

⌀102

6-3 SINGLE-VIEW DRAWINGS

Simple objects made of flat sheet or plate stock are often drawn as single-view drawings. Though it may seem like a contradiction to classify one-view drawings as multiviews, the classification is appropriate when they are drawn as flat, two-dimensional views. Chapter 4 covered single-view pictorial drawings, but single-view multiview drawings are different. They do not attempt to show the depth of the object; instead, the third dimension is given in a note or by written dimension. The purpose for using single-view multiviews is, of course, simplification. They are used whenever a second view would show only a minimum of information—that is, information that can be communicated better by other means. Figure 6-1 shows an example of a single-view drawing that uses a note to indicate the thickness.

In drafting, it is important to follow a logical set of procedures to produce a readable and accurate drawing in the shortest amount of time. The following procedures should be used for making single-view drawings. These steps are illustrated in Figure 6-2. Remember that all drawing construction work is done with very light line work.

1. Study the object or sketch of the object to be drawn. Determine what the major dimensions are and to what scale (full-size, 1/2 size, or whatever) it should be drawn.

Figure 6-2 The steps in constructing a single-view drawing. Detailed dimensions are omitted for clarity.

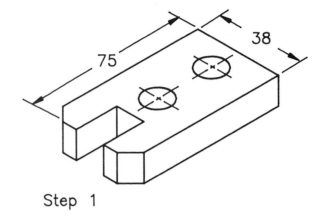

Step 1

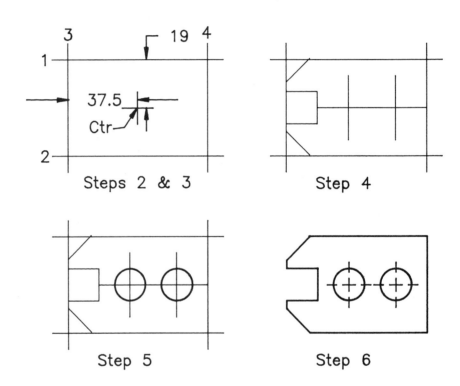

Steps 2 & 3

Step 4

Step 5

Step 6

2. Find the approximate center of the drawing sheet and establish points one-half the length of the object and one-half the width from that center.

3. Draw construction lines 1 and 2, making them 38 mm apart and parallel to each other and the edge of the paper. Draw lines 3 and 4, making them 75 mm apart and perpendicular to lines 1 and 2.

4. Measure and lay out the details of the object (the notch, cutoff corners, and holes). Use only centerlines to place the holes.

5. Draw circles (or arcs) with dark object lines.

6. Darken all straight lines, add centerlines, and erase any construction lines that appear too dark.

7. Add required lettering, borders, and so forth.

Note that nearly the entire drawing is done with construction lines before any dark lines are drawn. This seven-step process is nearly the same as the basic techniques described in Section 2-7.

6-4 TWO-VIEW DRAWINGS

Complex objects require more views. If the depth dimensions cannot be indicated by a simple note, then at least a second view is required. Figure 6-3 shows an example of a two-view drawing. Notice that both views are needed because neither gives a complete description of the object by itself—the features shown in the right side view would be difficult to describe in a note.

The steps involved in constructing two-view drawings are similar to those used for single-view drawings. The principal difference comes from having to plan for two views. With a single-view drawing, the choice of views is usually obvious, but not so with a two-view drawing. The object is more complex than one needing only a single view, and care must be taken to select views that will show all the necessary information for a complete understanding of the object.

Figure 6-4 shows several examples of the view selection process. In Figure 6-4a, the front and top views would probably be selected because the right side view (or any other view) does not contain any information not found in the front and top views. In Figure 6-4b, you could eliminate the front view because it is identical to the top view. In Figure 6-4c, the choice would be

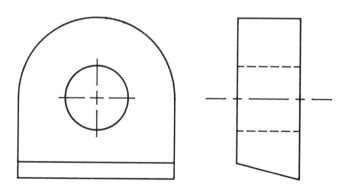

Figure 6-3 A typical two-view multiview drawing.

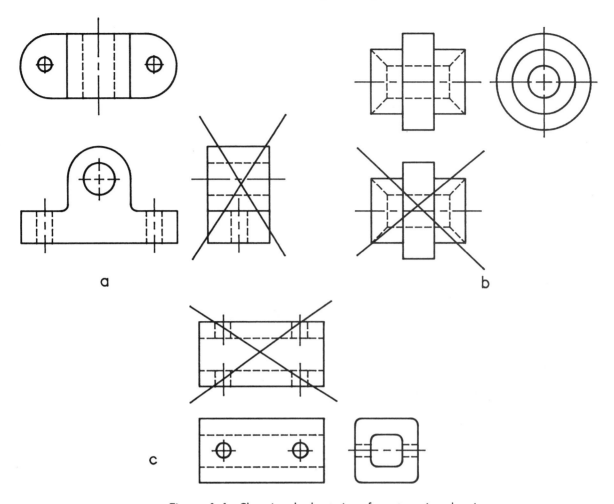

Figure 6-4 Choosing the best views for a two-view drawing.

the front and right side views because the top view shows the same information, but not as clearly. Remember, you should draw the minimum number of views that fully describe the object.

In establishing a two-view layout on your paper, you must always be sure that adjacent views are properly aligned and positioned and that the space between them is adequate. Alignment and positioning of views is extremely important. Refer to Figure 6-5 and take particular note of the positioning and alignment of the views. The positioning of the views is such that the top view is always drawn directly above the front view, the right side view is drawn to the right of the front view, and the left side view is always on the left. (The right side view could appear to the right of the top view, and the drafter can choose between the two positions.) Also notice that the views are aligned with each other. This means that every point in a view lines up with the same point

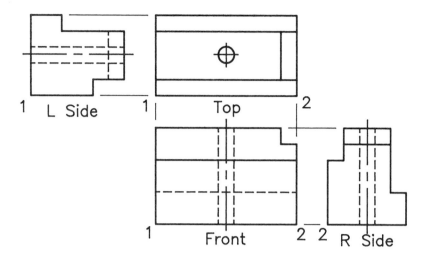

Figure 6-5 Four views showing positioning and alignment of the views. Note the alternate position of the left side view.

in adjacent views. In Figure 6-5 point 1 in the front view lines up with point 1 in the top view and point 1 in the left side view; point 2 in the right side view lines up with point 2 in the front view and point 2 in the top view. Proper positioning and alignment of views is necessary to ensure readability of multiview drawings. The easiest and surest way to accomplish this is to lay out and draw both views together, almost as if they were a single view.

The following procedures should be followed in constructing a two-view drawing. These steps are illustrated in Figure 6-6.

1. Determine which views to draw and what scale is appropriate for the object's size and detail.

2. Determine a starting point for the picture on your drawing sheet. Make sure that you have sufficient room for both views. If not, use a smaller scale or a larger sheet.

3. In both views lay out view outlines together by drawing construction lines that match the longest outside dimensions of the object.

4. Lay out details of the object, still using construction lines. It is necessary to mark only the centers of circles and arcs.

5. Draw circles and arcs with dark lines.

6. Darken all straight lines, add centerlines, and erase any construction line extensions that appear too dark.

7. Add required lettering, borders, and so on.

Figure 6-6 The steps in constructing a two-view drawing.

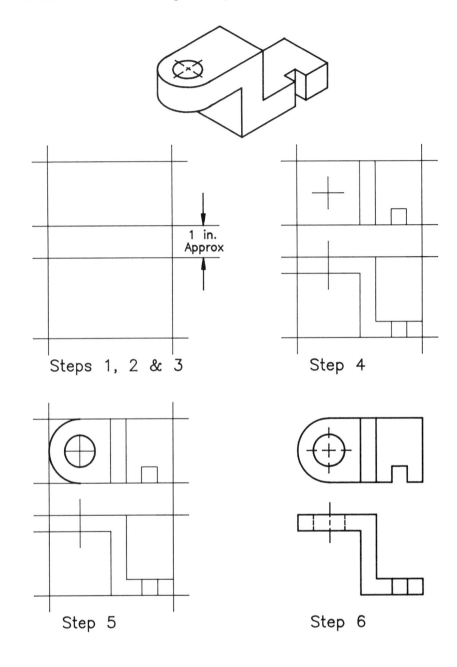

Steps 1, 2 & 3

1 in. Approx

Step 4

Step 5

Step 6

6-5 THREE-VIEW DRAWINGS

Three-view drawings are commonly used multiviews because many objects are too complex to be adequately described in one or two views. Figure 6-7 shows a typical three-view drawing. Try to visualize the object by using any two of the views and blocking out the third one. You will note that important information in each view is needed for a complete description.

In order to create three-view drawings, you must distinguish between related and adjacent views and the dimensions that are common to them. Fig-

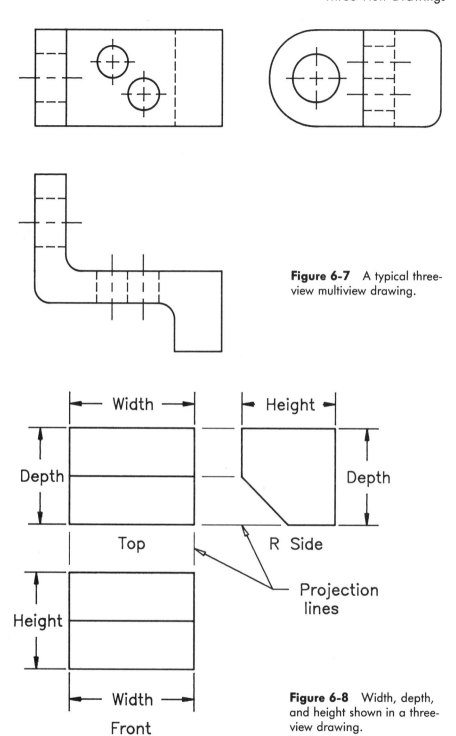

Figure 6-7 A typical three-view multiview drawing.

Figure 6-8 Width, depth, and height shown in a three-view drawing.

ure 6-8 shows a three-view drawing and indicates the principal dimensions observable in each view. In the top view the width and depth are seen, and in the front view the width and height are seen; the right side elevation view shows the depth and height. You can see that the width dimension is common to both the top and front views. Since these views are aligned and next to each other, it is easy to draw parallel construction lines between the views, as illustrated. The parallel construction lines are called **projection lines,** and the views are said to be **adjacent** to each other. Likewise, the top and right side views are adjacent, have a dimension (depth) in common, and can be connected by projection lines. The front and right side views have the height dimension in common, but they are not adjacent and have no projection lines between them. They are called **related** views because they are related through the front view.

Other terms could have been used for the principal dimensions, such as length or thickness. Though there is no set standard for which terms to use, it is generally agreed that certain terms are used for certain kinds of objects. For some objects, such as that shown in Figure 6-8, width, depth, and height are appropriate. For items such as boats or automobiles, length and width and height are appropriate. For objects that are uniformly flat (such as a piece of paper or sheet metal) and have one dimension that is smaller than the others,

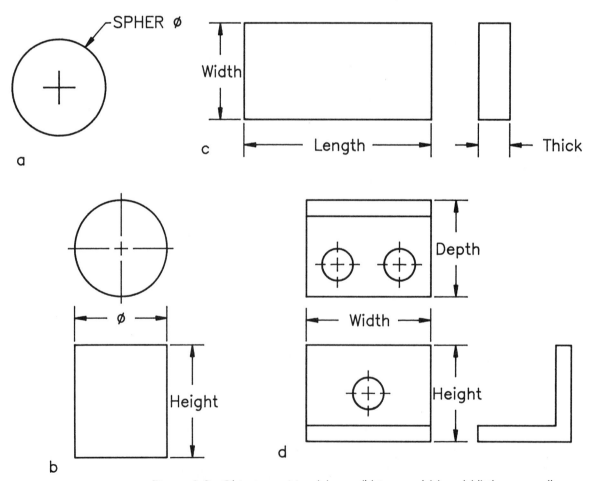

Figure 6-9 Objects requiring (a) one, (b) two, and (c) and (d) three overall dimensions.

length, width, and thickness are used. Some objects, such as cylinders, require two dimensions only. They are described by their diameter and length (unless, like a telephone pole, they are standing vertically, in which case diameter and height are used). One geometric shape, the sphere, requires only one dimension, its diameter. In the examples shown in Figure 6-9, notice that, usually, the number of dimensions closely corresponds with the number of views necessary to fully describe the object.

As noted in Section 6-4, laying out two adjacent views is relatively easy if you construct them both at the same time. Drawing the third view is not so easy, primarily because you must be certain to draw the common dimensions the same length in related views but you have no direct projection lines to help you. There are two methods for overcoming this problem: transferring dimensions and using 45° miter lines. The following list describes how to construct a three-view drawing by transferring dimensions. Use these procedures to construct a right side view as the third view, which is adjacent to the front view. These steps are illustrated in Figure 6-10.

1. Following the instructions in Section 6-4, construct a two-view drawing, but do not darken any lines.

2. Draw parallel projection lines from all points in the front view into the location for the right side view. These projection lines must be perpendicular to the top view–front view projection lines.

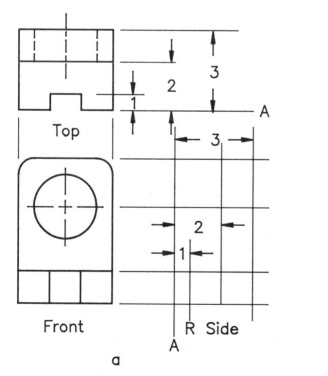

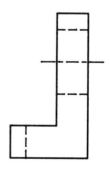

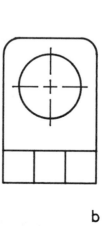

Figure 6-10 (a) Constructing the third view by projection and then transferring distances. (b) The completed drawing.

3. Identify the front face of the object in the top view, and establish it as a reference line. In Figure 6-10a the reference line is line A. Draw line A in the right side view about 1 in. from the front view and perpendicular to the projection lines.

4. In the top view, use dividers, a compass, or a scale to measure the distances 1, 2, and 3 from line A. Transfer the distances to the right side view. All measurements must be perpendicular to line A in both views.

5. Determine which lines are visible and which are hidden. After adding circles and arcs, darken the straight lines. Add centerlines.

6. Add required lettering, borders, and the like.

Almost the same procedure is followed to construct a three-view drawing with the right side view adjacent to the top view. Figure 6-11 illustrates the procedure.

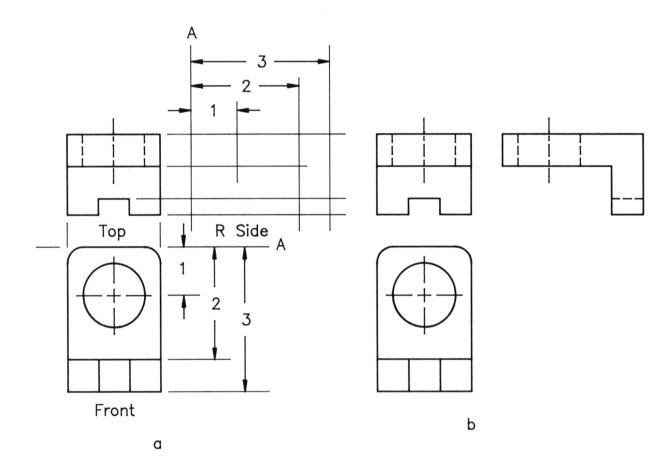

Figure 6-11 (a) Constructing the third view adjacent to the top view by projecting and then transferring distances. (b) The completed drawing.

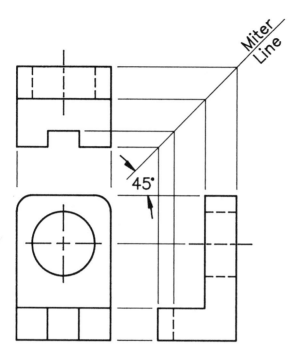

Except for steps 3 and 4, the steps for constructing a three-view drawing by using a 45° miter line are the same as the procedure that uses dividers. The steps in the miter method follow and are illustrated in Figure 6-12.

1. Following the instructions in Section 6-4, construct a two-view drawing, but do not darken any lines.

2. Draw parallel projection lines from all points in the front view into the location for the right side view. These projection lines must be perpendicular to the top view–front view projection lines.

3. Construct a line that starts at the top right-hand corner of the front view and extends up to the right at 45° to the projection lines.

4. Draw construction lines to the right from all points in the top view. These lines must be perpendicular to the front view–top view projection lines and must intersect the 45° miter line. From the intersections draw vertical construction lines downward to intersect the projection lines from the front view. This establishes all points in the right side view.

5. Determine which lines are visible and which are hidden. After adding circles and arcs, darken the straight lines. Add centerlines.

6. Add required lettering, borders, and the like.

A method for projecting the third view, called cross projection, is shown in Figure 6-13. In this example the front and right side views have been constructed first, as shown in Figure 6-13a. Projection lines from both views have been drawn into the area for the top view. The intersections thus formed establish all needed points for drawing the third view; you don't need to transfer any dimensions. Notice that the projection lines for the hole establish the limits of the ellipse created where the hole intersects the inclined surface in the top view. Figure 6-13b shows the completed drawing.

Figure 6-14 shows two methods for constructing a third view that includes a curved line that is not an arc or an ellipse. Figure 6-14a illustrates cross projection; Figure 6-14b employs projection and dimension transferring. The points established by either method are connected by using an irregular curve as described in Section 2-3.

Figure 6-13 (a) Using cross projection to construct a top view. (b) The completed views.

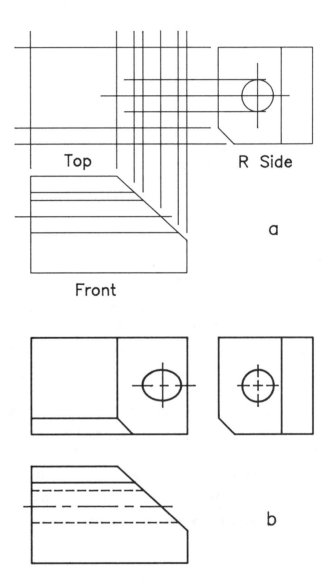

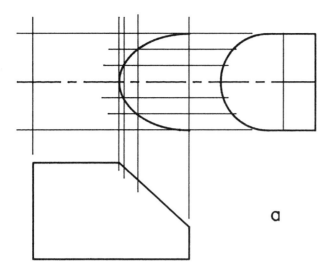

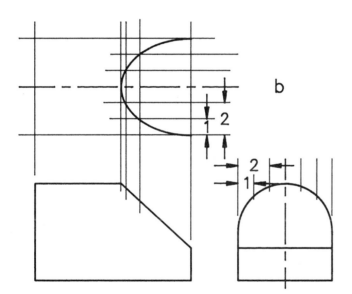

Figure 6-14   Using cross projection to construct a curved line. (b) Using projection and dimension transferring.

6-6 PARTIAL VIEWS

Drawing very complex objects often results in very complex and difficult-to-understand multiview drawings. To clarify these complex drawings, **partial views** are often used. These are views in which unnecessary details are omitted by showing only part of the object. Of course, drawing partial views not only simplifies the drawing, but also shortens the drafting time. There are two principal ways of simplifying views: by omitting confusing or misleading lines (especially hidden lines) and by omitting foreshortened surfaces.

In Figure 6-15b, both the right and left side views are simplified partial views. The left end of the part is not shown in the right side view, and the right end is not shown in the left side view. This makes the drawing much more understandable than the drawing shown in Figure 6-15a, where the

Figure 6-15 (a) Complete, but confusing, side views. (b) The same object drawn with partial, but readable, side views.

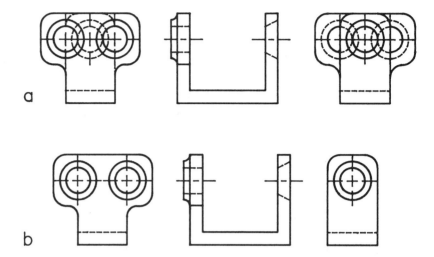

Figure 6-16 Partial views drawn using break lines to emphasize that the views are incomplete.

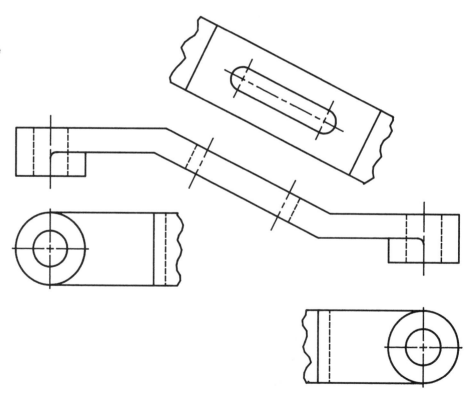

entire part is shown in all views. Which lines to omit in a partial view is a decision made by you, the drafter. There are no rules to follow, so you must use your judgment to decide what will make the drawing most understandable.

Where there is a chance that a partial view could be interpreted as a complete view, broken views are often used. These views are called broken views because they are drawn so that it appears as if part of the object were broken off. This is indicated by freehand lines called break lines (see Section 2-6). The broken line means that the object doesn't really look like the drawing, because the drawing is a partial view. Figure 6-16 illustrates several broken views of the same object. To project the entire object into each view would show an enormous amount of useless detail. The views would be more difficult to draw and more difficult to read. Breaks may be made anywhere and in any view.

In the case where a part must be drawn to a reduced scale (smaller than its actual size), important detail information often becomes difficult to read. This problem can be solved by drawing a partial detail view at an enlarged scale. Figure 6-17 shows an example. Note that the area to be enlarged is enclosed by a circular line and that the enlarged view has a title (DETAIL A).

Figure 6-18 shows another way of using broken lines in partial views. Whenever an object is very long in relation to its diameter or width and its cross section is constant, it is common practice to break the object in the middle and show only the ends. This permits the use of a larger scale and greater definition of detail.

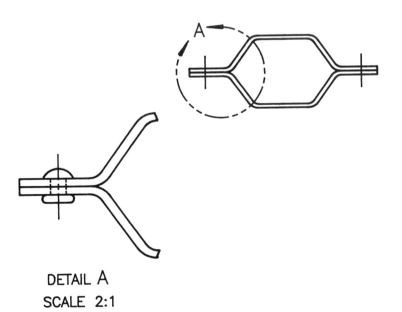

Figure 6-17 An enlarged partial view shows details.

DETAIL A
SCALE 2:1

Figure 6-18 Using breaks in partial views: conventional representations of long objects with constant cross sections.

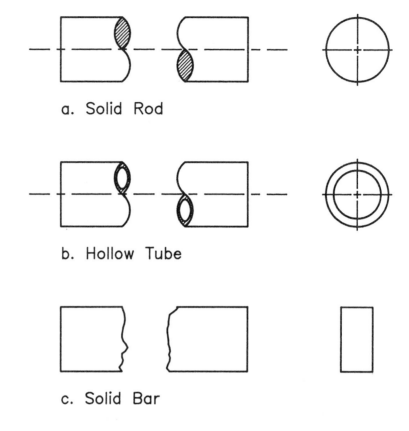

a. Solid Rod

b. Hollow Tube

c. Solid Bar

6-7 HIDDEN OBJECT LINE TECHNIQUES

Thus far in this chapter, only views with a minimum of hidden object lines have been presented. This has been done because it is good practice to select views for drawing that contain the greatest number of visible object lines and the least hidden object lines. Visible lines are nearly always much easier to read than hidden lines, as Figure 6-19 shows. The left side view gives much

Figure 6-19 Comparison of two views—the left side is much easier to read than the right.

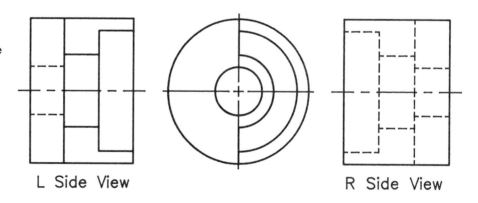

L Side View   R Side View

clearer information than the right side view. However, views of real objects usually contain a few hidden lines, and often contain a great many.

To make drawings with hidden object lines as legible as possible, certain techniques involving the use of hidden lines should be followed. Figure 6-20 illustrates these techniques.

- There should be no gap where a hidden object line intersects a visible object line unless the hidden line is a continuation of the visible line. (See points 1 and 3 in DETAIL A of Figure 6-20.)

- Hidden object lines that actually intersect each other should be joined at the apparent intersection. (See points 2 in DETAIL A and 5 in DETAIL B.) However, if they do not actually intersect, then a gap should be left. (See point 6 in DETAIL C.)

- Parallel hidden object lines should have staggered dashes, as shown in DETAIL C.

Following these techniques will significantly improve the readability of your drawings.

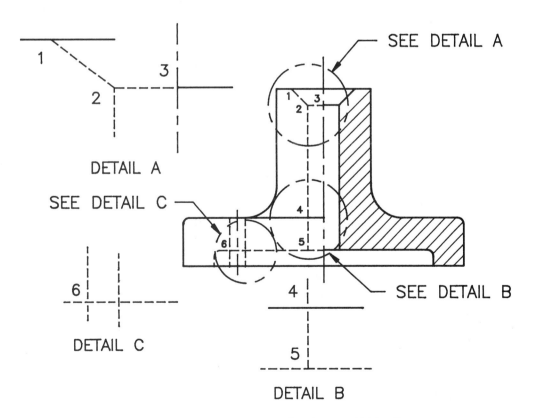

Figure 6-20 Detail views illustrating conventions for drawing hidden lines.

6-8 CONVENTIONAL REPRESENTATIONS

Certain views of objects, when drawn according to the multiview principles of Chapter 5, will not always appear as you might imagine they should. This sometimes leads to problems in reading the drawings and misinterpretation of their meanings. To prevent this, conventional representations are often used to create more legible drawings. Conventional representations are drawings of features and views that violate some of the principles of multiview drawing but provide a more meaningful picture as a result. The two most common conventional representations are rounded intersections and runouts.

Rounded Intersections

As noted in Section 5-4, surfaces adjacent to each other meet at their line of intersection, resulting in object lines. If, however, the intersections are rounded or radiused, there are no real intersections; no object lines should be drawn. As you can see in Figure 6-21, true representations of rounded intersections result in no object lines in the top view. This is misleading because the part appears to have a single surface with no change in direction. Actually, the part consists of three separate surfaces.

To clarify situations such as this, abrupt changes in surface direction are represented by phantom lines (see Section 2-6) drawn at the approximate intersection of the surfaces. This creates a much more informative drawing. Figure 6-22 shows several drawings that include phantom lines. Note that the external radii are called **rounds,** and the internal radii are called **fillets.**

Runouts

Whenever a curved surface is tangent to any other surface, another possibly misleading situation exists. The two surfaces meet with no actual intersection, and all lines simply end or fade out. This condition is called a **runout.** Runouts are indicated by lines ending with arcs, curving either outward or inward (depending on the shape of the cross section of the object), and ending at the tangent line. Examples are shown in Figure 6-23.

6-9 CONCLUSION

This chapter has covered the basic techniques for constructing one-, two-, and three-view multiview drawings. Partial views, the correct use of hidden object lines, and two commonly used drawing conventions have also been presented.

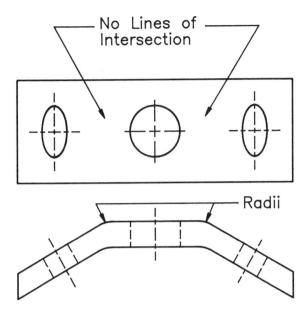

Figure 6-21 Drawn as a true projection, the top view gives a misleading impression of the object's shape.

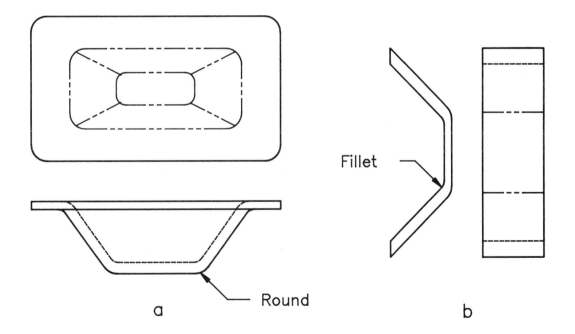

Figure 6-22 Phantom lines used to represent filleted and rounded corners.

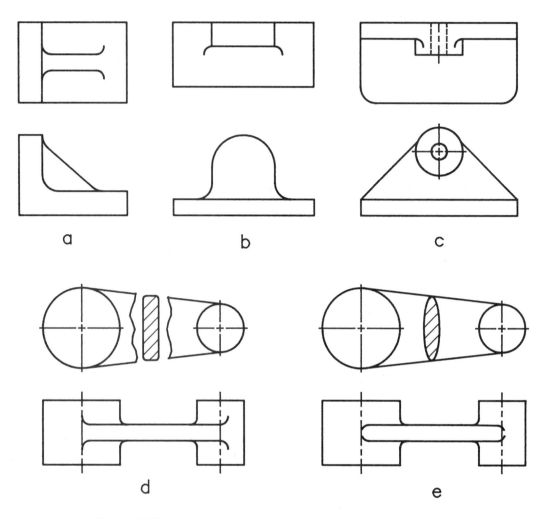

Figure 6-23 Conventional representations of runouts.

Though the examples and explanations in this chapter have been limited to drawings with one, two, or three views, note that three is not the maximum number of views that can be drawn. Complex objects may require four, five, or more views to present a complete and accurate description. Figure 6-24 shows an example of a four-view drawing that also illustrates the use of partial views and the correct conventions for showing runouts and hidden lines.

Check your general understanding of this chapter by answering the following questions.

REVIEW QUESTIONS

1. The choice of how many views to draw depends on which of the following?

    a. need for clarity   c. need for accuracy

    b. need for simplicity   d. all the above

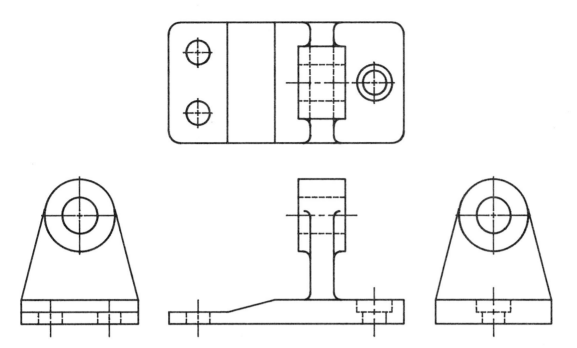

Figure 6-24 A four-view drawing. The left and right side views are partial views.

2. Which of the following statements is true concerning a single-view drawing?

   a. The third dimension is indicated in a note.

   b. Objects drawn have only two dimensions.

   c. They are used when there is no room on the drawing sheet for more views.

   d. Single-view drawings shouldn't be drawn.

3. The readability of a two- or three-view drawing is particularly dependent on:

   a. none of the below

   b. the alignment and positioning of the views

   c. always using the front view as the principal view

   d. always using the top view as the principal view

4. Which of the following statements is true concerning the dimensions of an object?

   a. Three dimensions are required to describe all three-dimensional objects.

   b. For general objects, length, width, and thickness are the most commonly used dimensions.

   c. At least one type of object requires only a single dimension.

   d. There is no relationship between the number of views and the number of dimensions required to describe an object.

5. Views with projection lines connecting them are called:

   a. related views   c. connected views

   b. adjacent views   d. projected views

6. Views that do *not* have projection line connections but do have one dimension in common are called:

   a. related views

   b. adjacent views

   c. unconnected views

   d. nonprojected views

7. In views connected by projection lines, how many dimension(s) are the same?

   a. varies from view to view

   b. one

   c. two

   d. three

8. Partial views are useful for:

   a. simplifying complex views

   b. complicating simple views

   c. saving drawing space

   d. none of the above

9. An actual (true) distance of 20 mm would be _____ on a drawing done to 1/4 scale.

   a. 20 mm

   b. 80 mm

   c. 5 mm

   d. 4 mm

10. Which of the following statements is true concerning hidden object lines?

   a. They should never be shown intersecting.

   b. They take precedence over all other lines.

   c. They have the same meanings as visible object lines.

   d. Gaps are always left between visible and hidden object lines.

11. Radiused intersections where no actual line of intersection exists are shown by means of:

   a. visible object lines

   b. hidden object lines

   c. centerlines

   d. phantom lines

12. When a curved surface is tangent to another surface the result is called:

   a. a fillet

   b. a runout

   c. a round

   d. a runin

PROBLEMS

6-1   Complete the three-view drawing of each object by drawing the view indicated by the bracket.

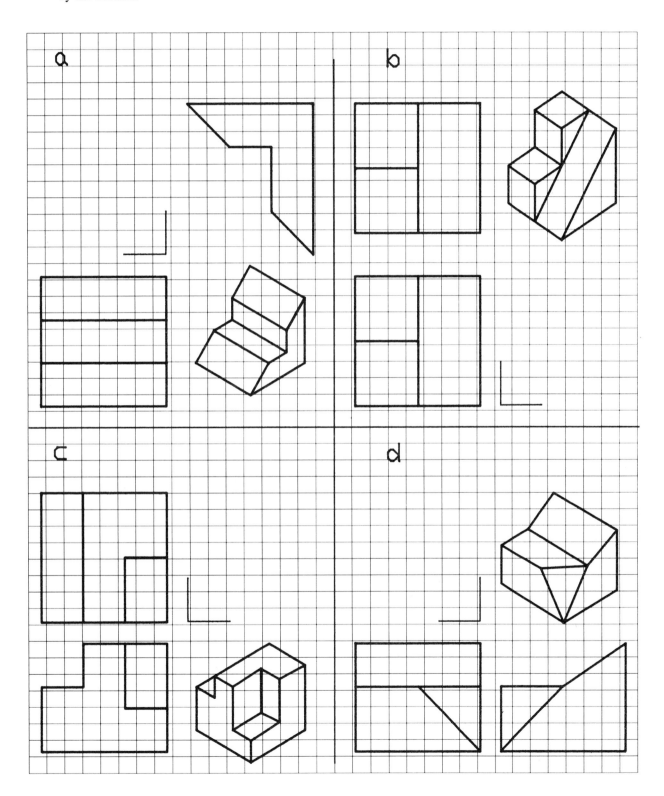

6-2   Complete the three-view drawing of each object by drawing the view indicated
by the bracket.

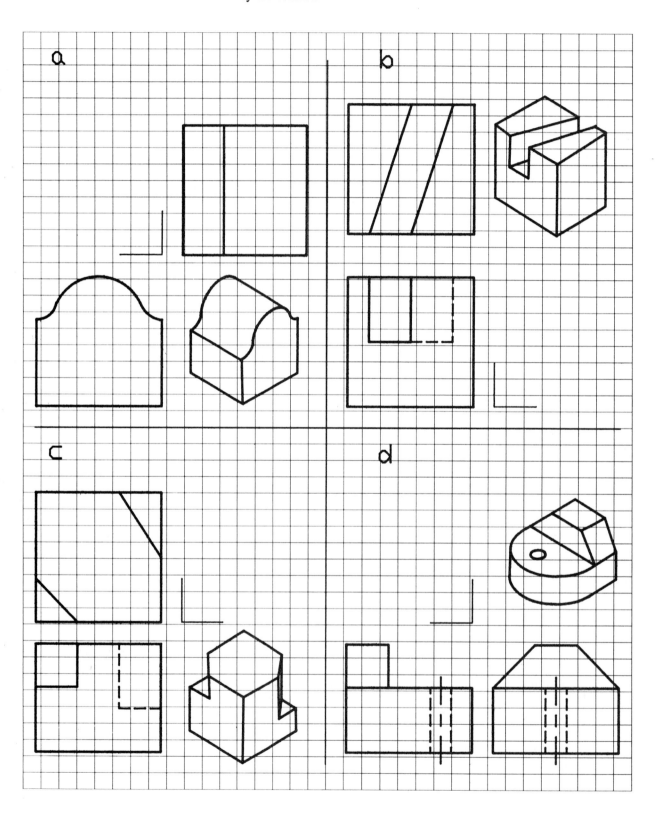

6-3   Complete the three-view drawing of each object by drawing the view indicated
by the bracket.

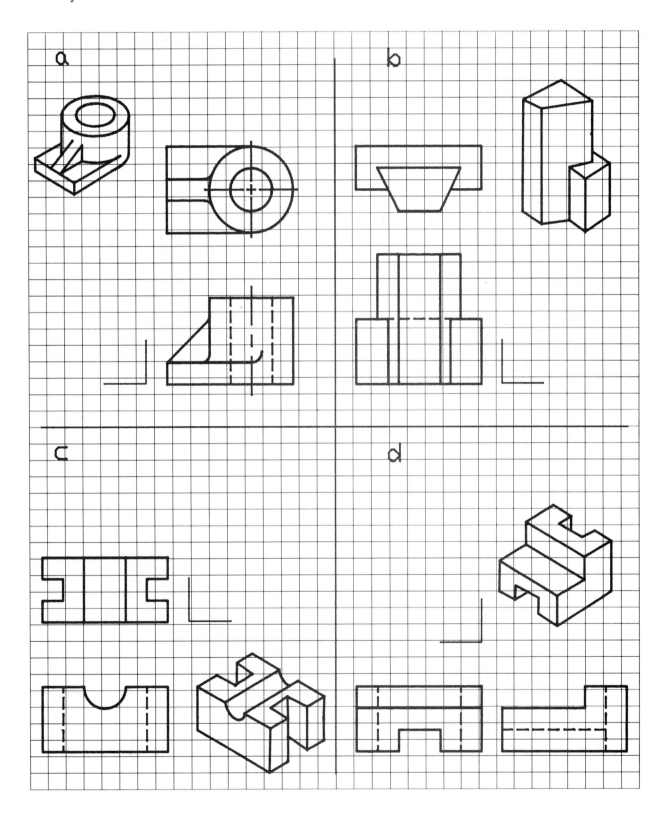

6-4 Complete the three-view drawing of each object by drawing the view indicated by the bracket.

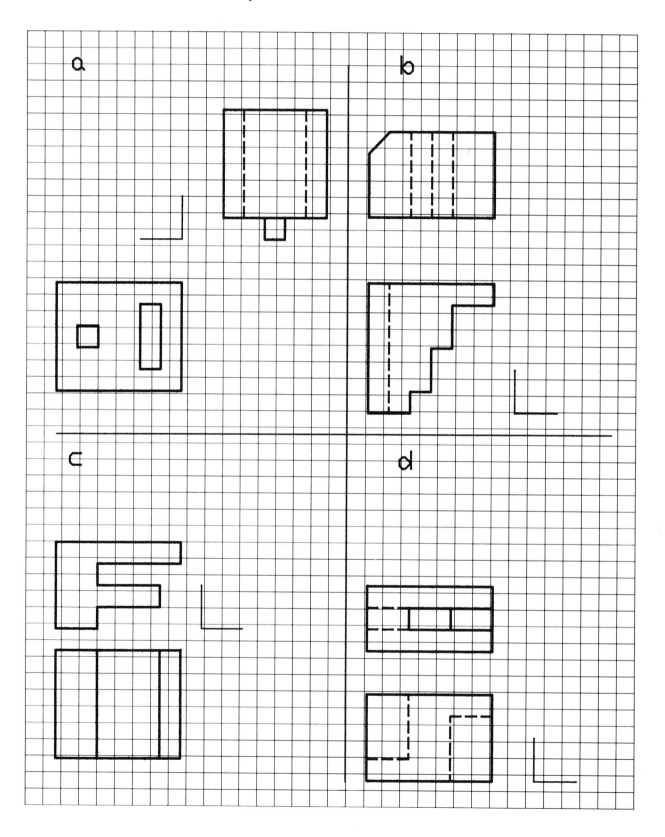

6-5 Complete the three-view drawing of each object by drawing the view indicated by the bracket.

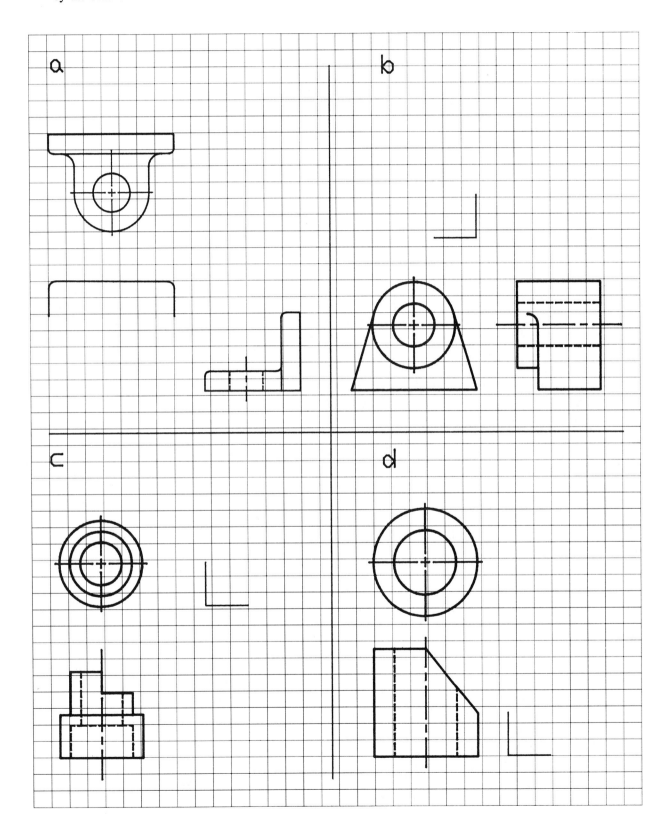

6-6   Complete the three-view drawing of each object by drawing the view indicated by the bracket.

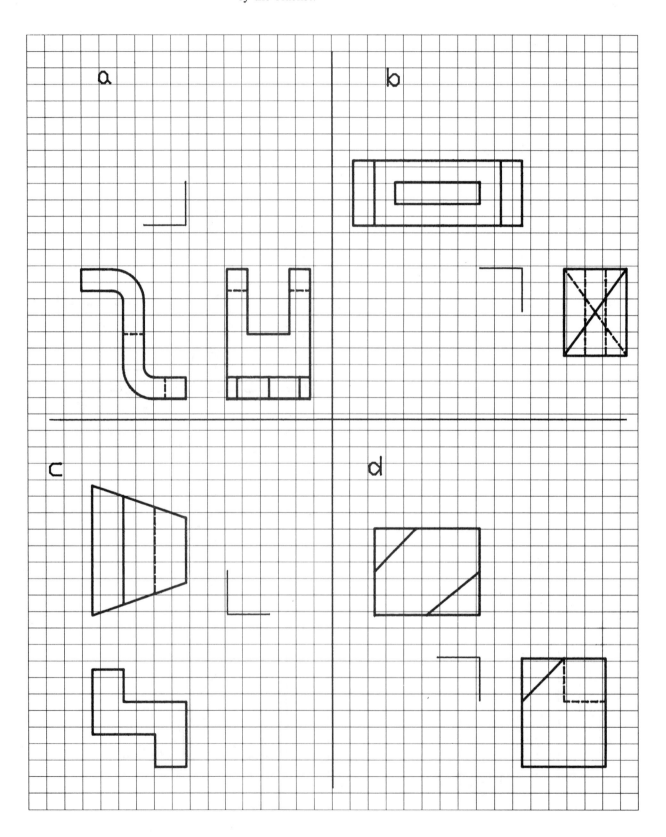

6-7 On graph paper draw a top, front, and left side view of the objects shown. The dimensions are given in spaces—one space is a small square on graph paper. Show all hidden lines in all views.

a

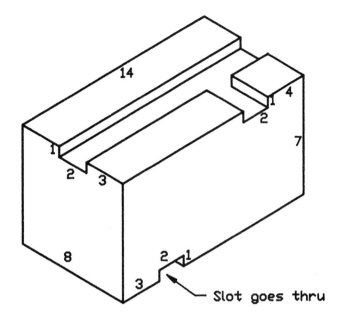

Slot goes thru

b

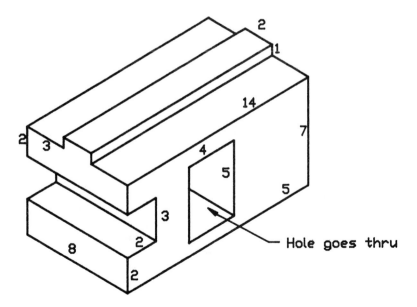

Hole goes thru

6-8 On graph paper draw a top, front, and left side view of the objects shown. The dimensions are given in spaces—one space is a small square on graph paper. Show all hidden lines in all views.

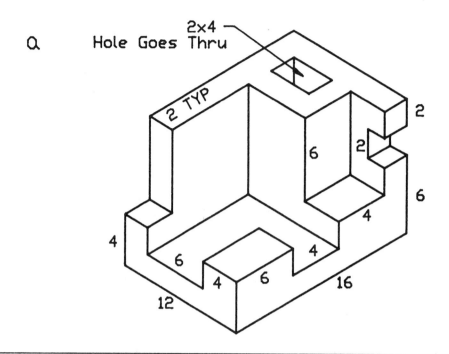

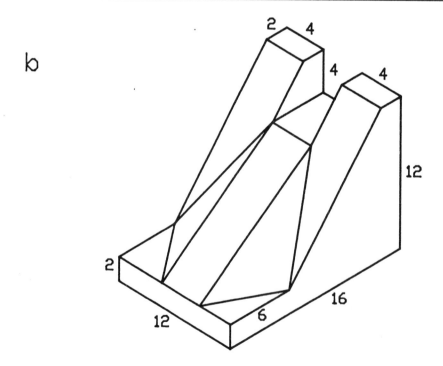

6-9   Make multiview drawings of each of the objects shown. Draw only those views necessary for a complete description. All dimensions are in millimeters. Draw to the scale required to fit each drawing on A4 (210-mm × 297-mm) paper.

 a

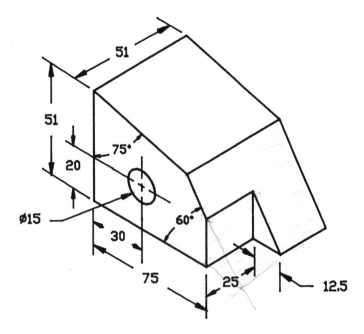

b

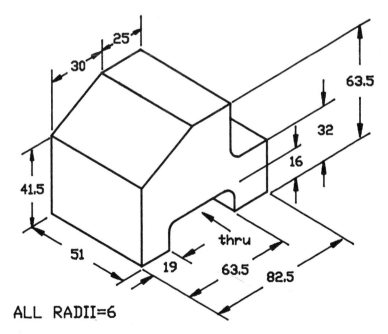

ALL RADII=6

6-10 Make multiview drawings of each of the objects shown. Draw only those views necessary for a complete description. All dimensions are in inches. Draw to the scale required to fit each drawing on A-size (8½-in. × 11-in.) paper.

a

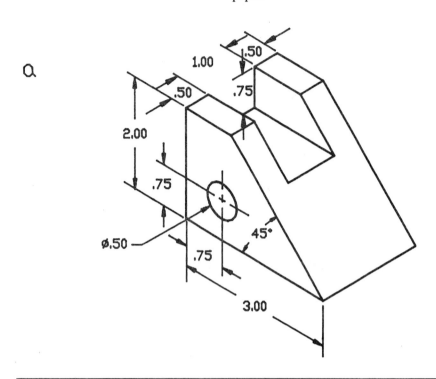

b

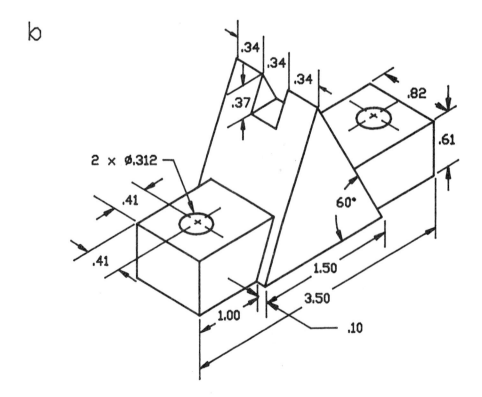

6-11   Make multiview drawings of each of the objects shown. Draw only those
views necessary for a complete description. All dimensions are in inches.
Draw to the scale required to fit each drawing on A-size (8½-in. × 11-in.)
paper.

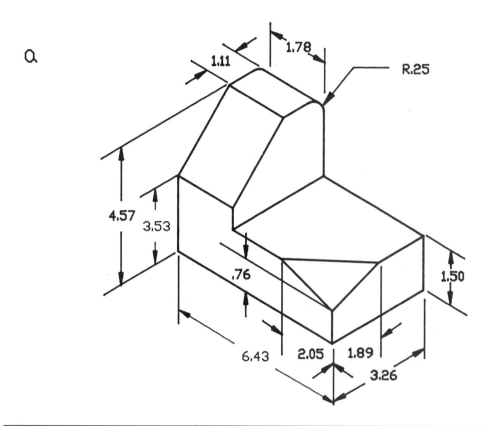

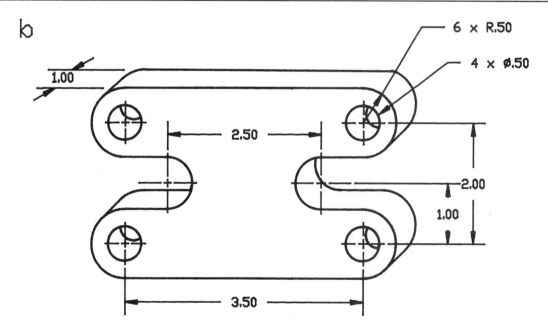

6-12 Make multiview drawings of each of the objects shown. Draw only those views necessary for a complete description. All dimensions are in inches. Draw to the scale required to fit each drawing on A-size (8½-in. × 11-in.) paper.

a

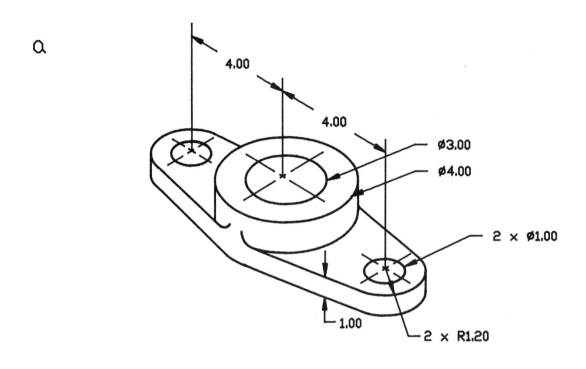

b

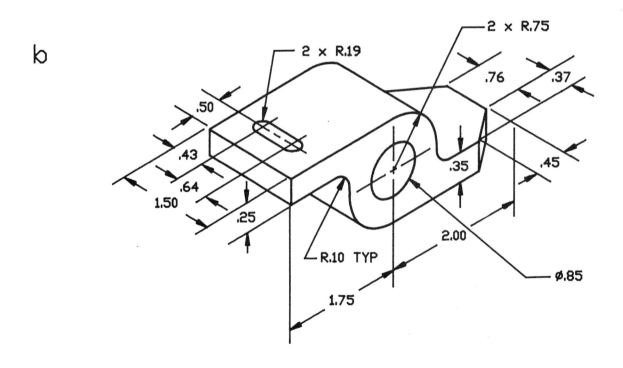

6-13 Make multiview drawings of each of the objects shown. Draw only those views necessary for a complete description. All dimensions are in millimeters. Draw to the scale required to fit each drawing on A4 (210-mm × 297-mm) paper.

a

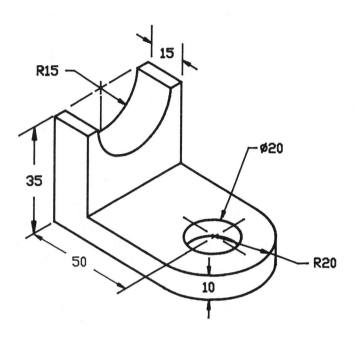

b

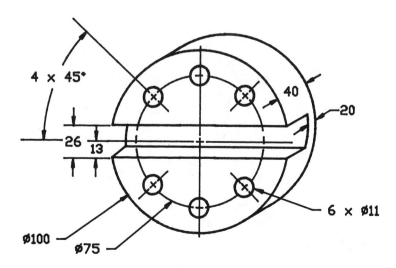

6-14 Make multiview drawings of each of the objects shown. Draw only those views necessary for a complete description. All dimensions are in millimeters. Draw to the scale required to fit each drawing on A4 (210-mm × 297-mm) paper.

a

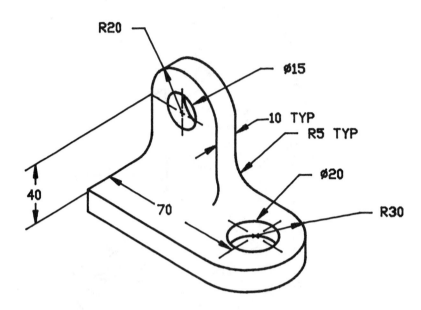

b

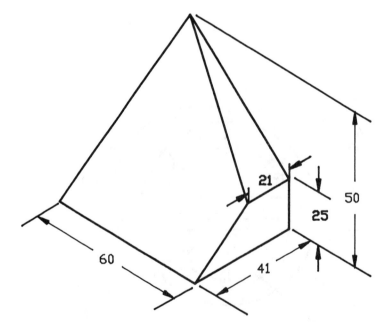

6-15   a. Redraw the objects in Problem 2-12, parts a, b, c, and d, as multiviews. Scale the figures for dimensions.

   b. Redraw the objects in Problem 4-1, parts a, b, c, and d, as multiviews. Scale the figures for dimensions.

   c. Redraw the objects in Problem 4-2, parts a, b, c, and d, as multiviews. Scale the figures for dimensions.

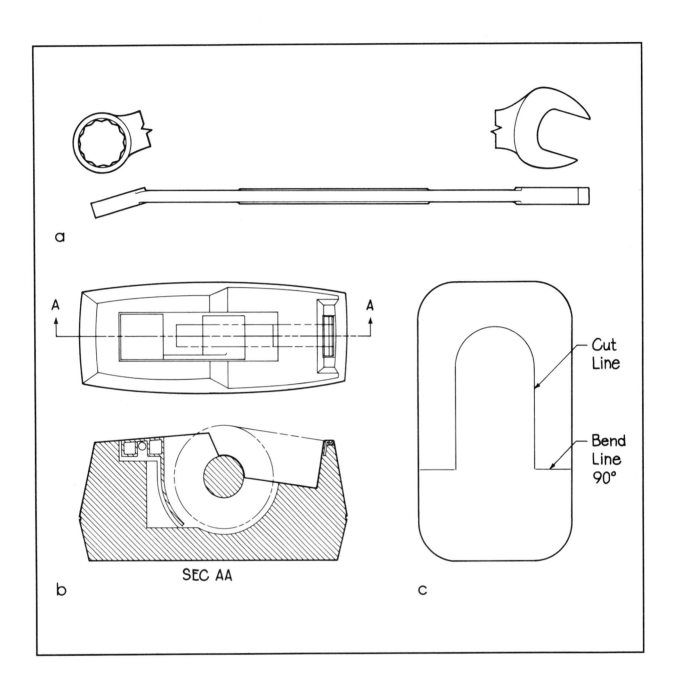

a

A ----A

SEC AA

b

Cut Line

Bend Line 90°

c

7

ADVANCED MULTIVIEW
TECHNIQUES

7-1 INTRODUCTION

The illustration that opens this chapter shows drawings of three complex objects. These drawings do not show just the standard two or three views; they make use of some advanced techniques to clearly describe these objects. Part a of the illustration shows a wrench described with one standard view and two partial auxiliary views. Part b shows a side view and a section view of a tape dispenser. The roll of tape is indicated by the phantom lines in both views. Part c is a development drawing showing the flat pattern of a common sheet-metal object. If it is cut and bent as called for, what will it be? You will find the answer at the end of the chapter.

There are many objects so complex and containing so many detailed features that they cannot be adequately described by using the standard type of multiview drawing. This chapter examines several advanced multiview drafting techniques that are used to simplify and clarify complex drawings. These include auxiliary views, section views, removed views, and developments. Auxiliary views are used to show the true size and shape of inclined and oblique surfaces. Section views are drawn when objects have complex interior detail that would be difficult to describe with hidden lines. Removed views are any views not drawn in alignment with their adjacent view. That is, they are removed to some other part of the drawing, to prevent crowding. Developments are drawings of objects made from sheet materials shown as flat patterns.

After you complete this chapter, you should:

1. understand the principles of projecting and drawing auxiliary views

2. know the difference between primary and secondary auxiliary views

3. be able to draw primary auxiliary views

4. understand the principles of section views

5. recognize the different types of section views and understand their purposes

6. be able to draw the basic types of section views

7. understand the concept of removed views

8. understand the principles of flat pattern development

215

7-2 PRIMARY AUXILIARY VIEWS

A surface must be parallel to a projection plane for it to appear true size and shape (TS&S). In other words, a TS&S surface must be perpendicular to the lines of sight because the lines of sight are always perpendicular to their projection planes in a multiview drawing. Chapter 5 showed that only normal surfaces appear TS&S when projected onto the planes of a standard projection box, because only normal surfaces are parallel to their projection planes. Inclined and oblique surfaces are not parallel to the standard projection planes; therefore, they cannot appear TS&S in any of the six standard views. However, the projection box can be modified and projection planes created that are parallel to inclined and oblique surfaces. These new projection planes result in views called **auxiliary views.** There are two kinds; primary auxiliary views, which show inclined surfaces TS&S, and secondary auxiliary views, which show oblique surfaces TS&S.

Primary Auxiliary View Principles

Figure 7-1a shows an object with an inclined surface in a standard projection box. Surfaces A and B of the object are normal, and they are parallel to the top and front projection planes, respectively. Surface C is inclined, is not parallel to any of the projection planes, and does not show TS&S in the top, front, or right side views. However, in Figure 7-1b, notice that lines 1,2 and 3,4 are

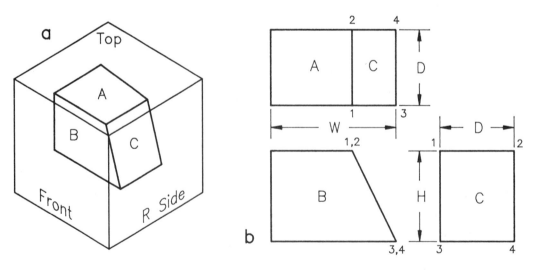

Figure 7-1 An object with an inclined surface in a standard projection box and a three-view drawing of the same object.

true length (TL) in the top view, that lines 1,3 and 2,4 are TL in the front view, and that these lines form the boundaries of surface C. This information can be used to create a TS&S view of surface C.

One way to obtain a TS&S view of surface C is by creating an auxiliary projection plane parallel to surface C. The modified projection box is shown in Figure 7-2. Notice that an auxiliary projection plane parallel to surface C is used instead of the right side projection plane. Surface C appears TS&S in the new auxiliary view because the lines of sight for that view are perpendicular to surface C.

To understand how the auxiliary view will be placed on the drawing sheet, imagine that the edge between the auxiliary projection plane and the front projection plane is the hinge line. The auxiliary plane "opens up" around this hinge and unfolds onto the plane of the drawing sheet, as shown in Figure 7-3.

When the box has opened and all projection planes have rotated onto the plane of the paper, the arrangement is as shown in Figure 7-4. The dimensions are almost the same as those in the standard three-view drawing shown in Figure 7-1. Depth and width are seen in the top view and height and width in the front view. Note the difference between the right side view and the auxiliary view, however. Both show the depth dimension, but the height does not appear in the auxiliary view as it does in the right side view.

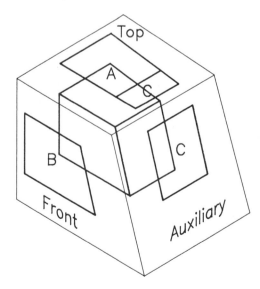

Figure 7-2 An auxiliary projection plane used in place of the right side projection plane.

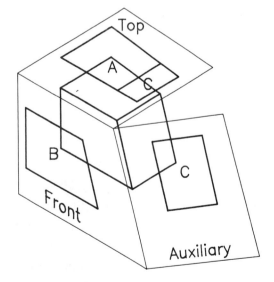

Figure 7-3 Unfolding the auxiliary projection plane.

Figure 7-4 A multiview drawing with an auxiliary view used instead of a right side view.

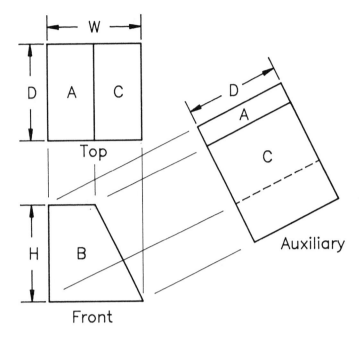

Primary Auxiliary View Construction

The steps in constructing a primary auxiliary drawing are nearly the same as those used to construct a standard three-view drawing (see Section 6-5). Figure 7-5 illustrates the procedures, and they are summarized in the list that follows. Note that, in drawing auxiliary views, it is standard practice to draw partial views showing only the inclined or oblique (non-TS&S) surface rather than the whole object. This practice makes the drawings both easier to draw and easier to read.

1. Starting with a standard multiview drawing with any number of views, identify inclined surfaces and locate an edge view for each one. Note that an oblique surface does not appear as an edge in any standard view. Figure 7-5 has one inclined surface, numbered 1,2,3,4.

2. Draw light projection lines perpendicular to the edge view of the inclined surface, from points 1,2, and 3,4 into the area where the auxiliary view is to be drawn.

3. Identify the front face of the object in the left side view and establish it as a reference line (coincident with line 1,3 and labeled "A" in Figure 7-5). Draw line A in the auxiliary view at a convenient distance from the front view. Reference line A must be parallel to the edge view of the inclined surface and perpendicular to the projection lines.

4. In the left side view, use dividers or compass to measure the distance D from line A and transfer it to the auxiliary view, thus locating points

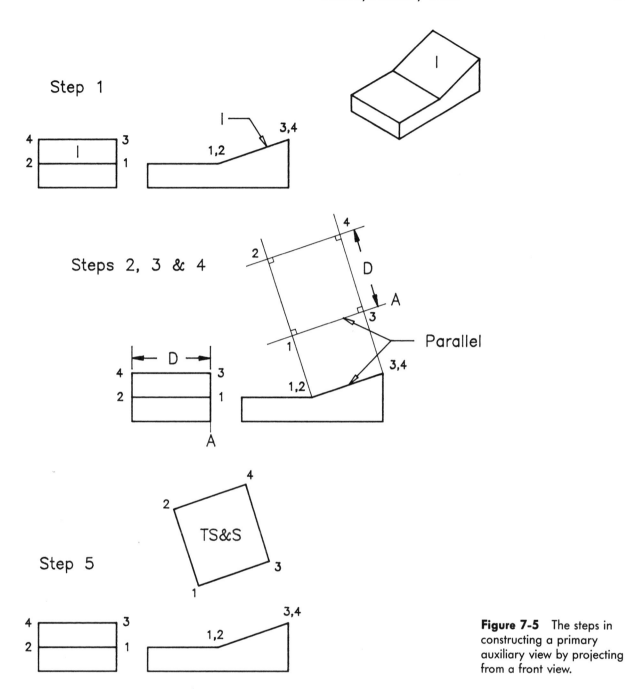

Figure 7-5 The steps in constructing a primary auxiliary view by projecting from a front view.

1, 2, 3, and 4. Notice that line 1,2 and line 3,4 are TL in the left side view and line 1,3 and line 2,4 are TL in the front view. In drawing the auxiliary view, lay out and connect these four TL lines to construct a TS&S view of surface 1,2,3,4.

5. Complete the view by darkening the object lines and erasing construction lines as needed.

In Figure 7-5 note several things in particular. Though the inclined surface shows TS&S in the auxiliary view, all the other surfaces would have appeared foreshortened had they been drawn. Notice the position of the auxiliary view relative to the standard views. Try to draw adjacent views within 20 mm to 30 mm (approximately 1 in.) of each other. With auxiliary views this is not always possible: Sometimes they will be very close and sometimes far apart. As a rule, keep views as close together as is practical but *never* overlap them.

Figure 7-6 shows another example of the construction of an auxiliary view that shows an inclined surface TS&S. Notice that the auxiliary view is a complete view showing all surfaces and hidden lines. The result is somewhat confusing. It is much better practice to follow the example in Figure 7-7, which shows partial primary auxiliary views used in place of side views.

7-3 SECONDARY AUXILIARY VIEWS

Secondary auxiliary views are needed to show oblique surfaces TS&S. The methods used for their construction are relatively complex and are the subject

Figure 7-6 A complete primary auxiliary view projected from a top view.

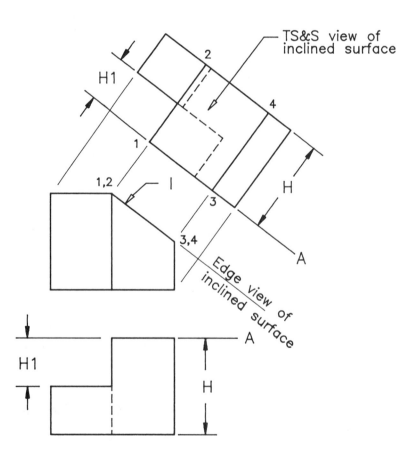

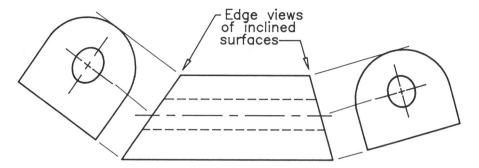

Figure 7-7   Two partial primary auxiliary views projected from a front view.

of more advanced texts. However, an example of the use of secondary auxiliary views is shown in Figure 7-8. Notice that a primary auxiliary view is drawn first to show the oblique surface as an edge. Then a secondary auxiliary view is projected from the edge view to show the oblique surface TS&S.

7-4 SECTION VIEWS

Some objects have very complex interior detail. Drawing these parts by using standard views often results in confusion because of the need to use numerous hidden lines. Sectional views, or **section views,** are used to show objects' interior detail without hidden lines, thus improving the clarity and accuracy of the drawing. Figure 7-9 shows an example of a complex part with a standard left side view and a right side section view. Compare the two for clarity and legibility.

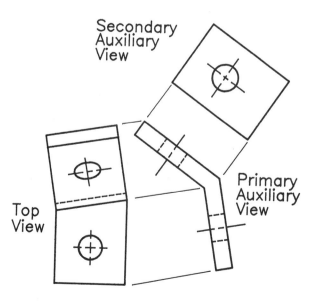

Secondary
Auxiliary
View

Primary
Auxiliary
View

Top
View

Figure 7-8   A secondary auxiliary view projected from a primary auxiliary view.

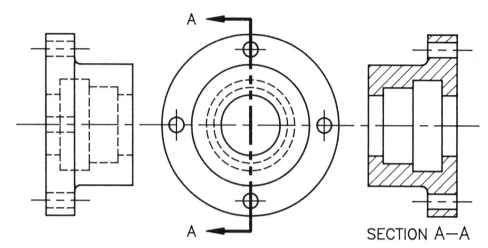

Figure 7-9   A comparison of a section view and a standard view.

Section View Principles

Section views are created by theoretically cutting the object and removing the portion between the observer and the cut, as shown in Figure 7-10, parts a and b. The interior detail, including all exposed surfaces and intersections, may then be shown with solid lines, as shown in Figure 7-10e, instead of hidden lines, as shown in Figure 7-10d. To further clarify the drawing, those surfaces that were cut are crosshatched with thin diagonal lines called section lining (see Figure 7-10e).

To make it clear where the section view was taken or projected from, a very heavy dashed line is shown passing through the view where the theoretical cut was made (see Figure 7-10c). This line represents the edge view of an image plane. But since it slices through the object, it is called a cutting plane. Arrows on the ends of the cutting plane line indicate the direction of sight for the section view. These cutting plane lines are also referred to as view indicators. Both cutting plane lines and section views are identified by letters or numerals 7 mm high (or .25 in. high) as shown in Figure 7-11. The cutting plane line may be omitted where its location is obvious.

Section views are drawn in a manner similar to that of standard views, with a few very important exceptions. In section views, only surfaces and edges that are either in or on the arrow side of the cutting plane are shown. All visible object lines, but few hidden object lines, are shown. Surfaces that are cut by the cutting plane are crosshatched. Surfaces not cut, such as the far side of a hole, are not crosshatched. Figure 7-11 shows the proper use of cross-hatching. The lines for cross-hatching are thin lines best drawn with a 45° triangle. Scribing a line on your triangle, parallel to one edge, will help you space the lines correctly.

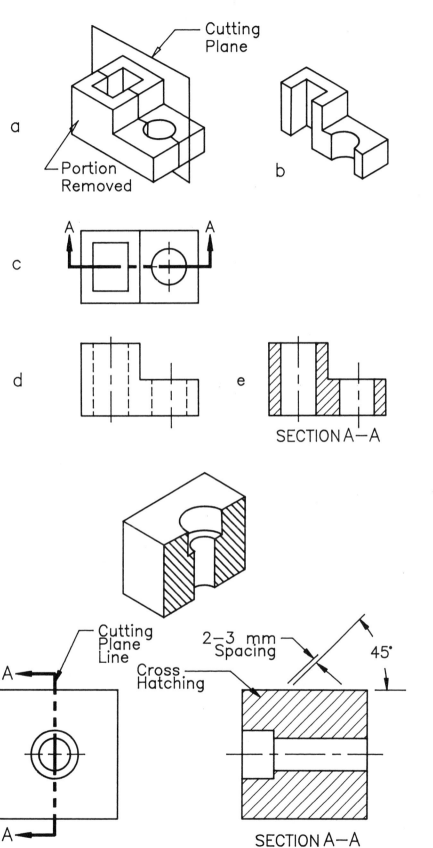

a

Cutting Plane

Portion Removed

b

c

A   A

d

e

SECTION A—A

Figure 7-10 (a) The cutting plane passing through the object. (b) The resulting section. (c) Standard top view showing cutting plane line. (d) Standard front view. (e) A section view replacing the front view.

Figure 7-11 Section view techniques, including cross-hatching.

Cutting Plane Line

A

A

Cross Hatching

2–3 mm Spacing

45°

SECTION A—A

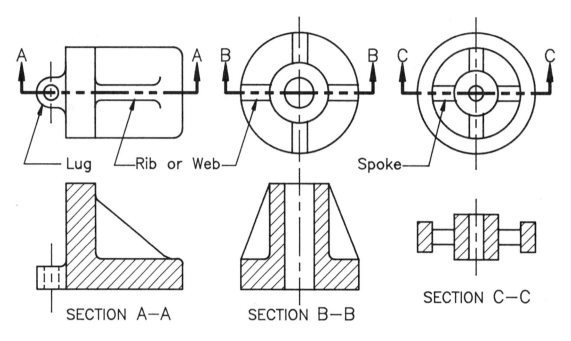

Figure 7-12   Ribs, webs, lugs, and spokes in section views.

Not all "cut" solid material needs to be crosshatched. Some features lying on the path of the cutting plane are solid and have no interior detail worth exposing. Other features are very thin in comparison to their surroundings. As Figure 7-12 shows, where a cutting plane passes through a rib, web, lug, spoke, or other feature thinner than the surrounding material, only the outline of that feature is drawn. This practice avoids giving a false impression of uniform thickness or solidity. In addition, it makes drawings clearer and avoids unnecessary work.

Many different types of section views are used for describing different types of objects and features. Descriptions of the most important types follow.

Full Sections

A full section is one in which the cutting plane passes fully through the object in a straight line. Full sections provide a view of the whole object and are the most commonly used type. Figure 7-13 shows a typical example. Notice that the notch at the back of the object does not appear in the section view because it is hidden; the notch in front does not appear because it is on the part of the object that has been removed.

Offset Sections

An offset section is a full section in which the cutting plane has been offset to pass through detail not lying in a straight line, thereby reducing the number

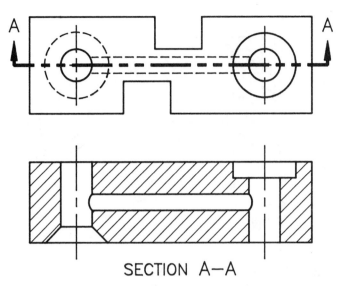

SECTION A—A

Figure 7-13 A full section view.

of section views needed. Figure 7-14 shows a typical example. Notice that no lines are shown in the section view where the cutting plane changes direction. In fact, the offset section view looks just like a full section view. The cutting plane may be bent any number of times, but all bends must make 90° angles.

Aligned Sections

A variation of the offset section is the aligned section, which is used to show partially symmetrical parts (see Figure 7-15). The cutting plane is bent and curved to pass through any features that need to be shown. The features are then rotated into alignment in a single plane and projected into the section view. This provides more clarity than a true projection, and it also makes the drafting task much easier. The cutting plane line is sometimes omitted in

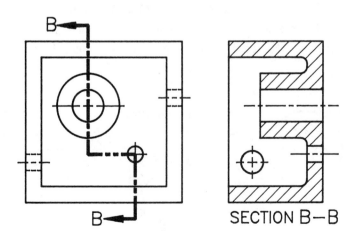

SECTION B—B

Figure 7-14 An offset section view.

Figure 7-15 An aligned
section view.

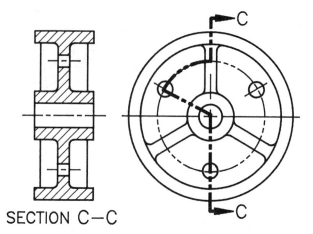

Figure 7-15 An aligned
section view.

aligned sections because it is usually obvious where the section view comes
from. This technique of drawing features in alignment is also used with standard views.

Revolved and Removed Sections

A revolved section is produced when a full section view is simply revolved
90° and drawn within the outline of the object. If the section view is drawn
outside the object outline, it is called a removed section. No cutting plane line
is shown, but a centerline indicates the location of the axis of revolution (see
Figures 7-16 and 7-17). The main views may be broken or not. Revolved sections are commonly used to show the shape of a long object with a changing
cross section.

Half Sections

A half-section is formed when a cutting plane is passed only halfway through
an object. It is most useful for showing cylindrical objects, which—if
symmetrical—need not be drawn in full section. An example is shown in Figure 7-18. Half of the side view is shown as a section view and the other half
as a standard view. Notice that only a centerline separates the standard view

Figure 7-16 Revolved
section views.

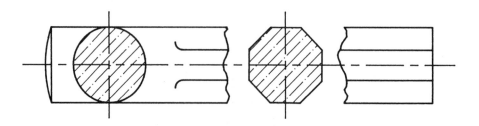

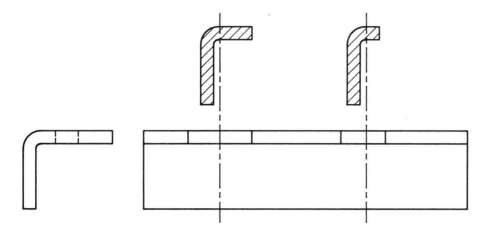

Figure 7-17 Removed section views.

half from the section view half. Hidden lines may or may not be shown in the nonsectioned half of the view. Cutting plane lines may be omitted because it is usually obvious where the section view comes from.

Broken-Out Sections

Another way of showing only part of an object in section is with a broken-out section. A portion of the object is removed, indicated by a freehand broken line, and that portion only is crosshatched. The remainder of the object is drawn as a standard view, as shown in Figure 7-19. No cutting planes are needed because the broken-out section identifies itself.

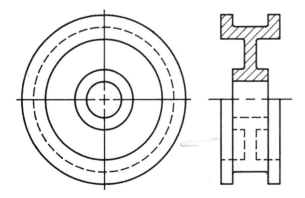

Figure 7-18 A half-section view.

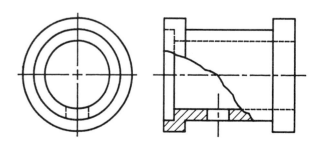

Figure 7-19 A broken-out section view.

7-5 REMOVED VIEWS

Often the complexity of an object requires a drawing with so many views that, if they were all projected with correct alignment, they would overlap each other. This could occur with standard views, auxiliary views, and section views. This difficulty is overcome by removing some of the views to other places on the drawing. Figure 7-17 showed examples of removed section views. A removed view is different. **A removed view** is any view not drawn in alignment with the view from which it is projected. It is necessary to show where each removed view came from; otherwise, the drawing may be unreadable. Showing where removed views come from is accomplished by using view indicators, which are similar to the cutting plane lines used for section views. View indicators are heavy, dashed lines with arrows pointing in the direction of the lines of sight for a particular view. Letters or numbers are placed beside the arrows and also under the removed view so that there is a cross-reference between them. Figure 7-20 shows some examples of removed views and view indicators. Notice that, although the removed view may be placed anywhere on the drawing, its attitude, or orientation, is the same as if it were a view drawn in proper alignment with its adjacent view.

7-6 DEVELOPMENTS

A drawing of the complete surface of an object flattened or laid out on a single plane is called a development, or a **flat pattern.** Developments are made for

Figure 7-20 Removed views and view indicators.

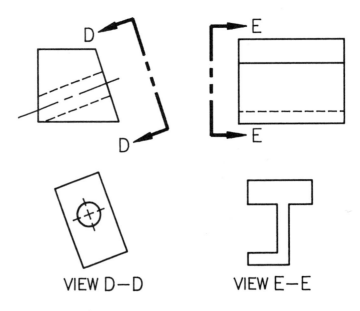

VIEW D—D   VIEW E—E

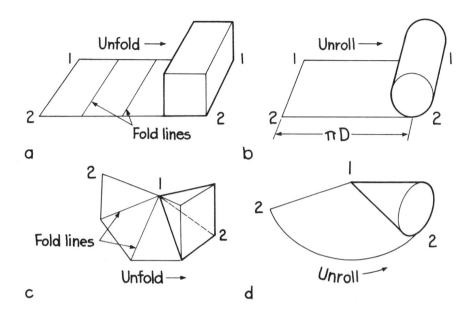

Figure 7-21 Developments of a (a) prism, (b) cylinder, (c) pyramid, and (d) cone.

objects made of sheet metal, cardboard, and most other materials used in sheet form. The resulting flat patterns are then folded or rolled to form objects such as boxes, stovepipe, and the like. The geometric forms that can be drawn as flat patterns are said to be developable and include prisms, cylinders, pyramids, and cones (see Section 3-5). Figure 7-21 shows these four forms and their developments.

Development of Prisms

Prisms are geometric forms that contain parallel, plane, lateral surfaces. Many of the objects shown in this text are prisms; an example is shown in Figure 7-22a. Imagine that the prism in the figure is hollow and that the six surfaces are thin sheets of metal or cardboard joined together at their edges. If you cut along some of the corners, but leave the remainder connected, the object can be unfolded as shown in Figure 7-22b. In Figure 7-22c, the object is shown unfolded completely on one flat plane. Note that the development shows every surface and that every surface is shown TS&S.

The list that follows presents the steps involved in developing a prism. The steps are illustrated in Figure 7-23.

1. Draw a multiview showing the TS&S of all surfaces.
2. Draw a horizontal line with a length equal to the base perimeter. This is called a stretch-out, or base line. On the multiview measure the lengths of line 1,2, line 2,9, line 9,10, and line 10,1 and transfer them to the stretch-out line. Mark points 1, 2, 10, and 9.

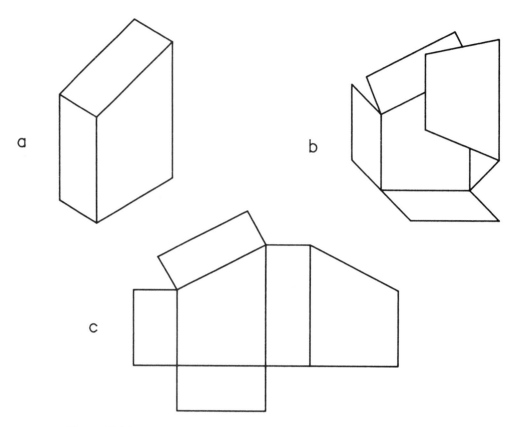

Figure 7-22   A prism seen as (a) a solid, (b) partially unfolded, and (c) completely unfolded.

3. Through points 1, 2, 9, and 10 on the stretch-out line draw four construction lines perpendicular to the stretch-out line. Measure and transfer from the multiview the distances from points 1 to 3, 2 to 4, 9 to 7, and 10 to 8. Mark points 3, 4, 7, and 8. Notice that the development starts at one of the shortest corners (1,3) and ends at the same corner.

4. Locate points 5 and 6 by swinging arcs of appropriate lengths from points 3 and 7 (for point 5) and from points 4 and 8 (for point 6). Connect all points as indicated in the multiview. This establishes the edges of the lateral surfaces.

5. Construct the top, bottom, and inclined surfaces from the multiview information. Notice that they are connected to lateral surfaces along their longest common edge.

6. Darken the appropriate lines, erase unnecessary marks, and the flat pattern is complete. Note that the lines separating surfaces represent not only the boundaries of the surfaces, but also the fold lines around

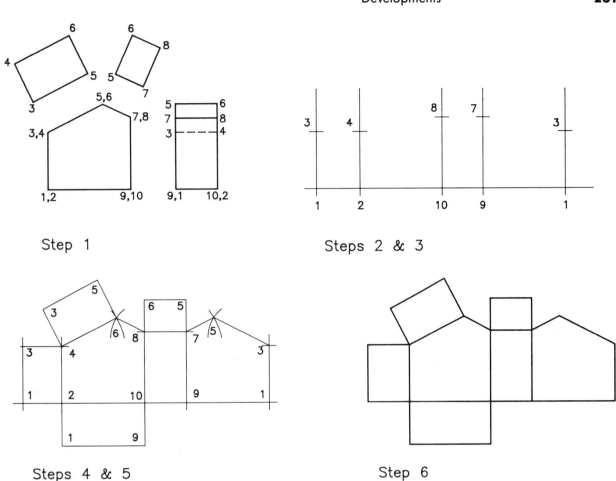

Figure 7-23 Steps in developing a flat pattern of a prism.

which the flat pattern will be bent to form the three-dimensional object.

Development of Cylinders

Cylinders are developed in a fashion similar to that of prisms. The list that follows presents the steps involved in developing a cylinder. The steps are illustrated in Figure 7-24.

1. Draw a multiview showing the height and diameter of the cylinder.

2. Divide the circumference (as shown in the top view) into 12 equal spaces. You may do this by measuring equal angles of 30° each (30 * 12 = 360°) or by striking arcs equal to the radius (R) as shown in detail A. Project these points into the front view.

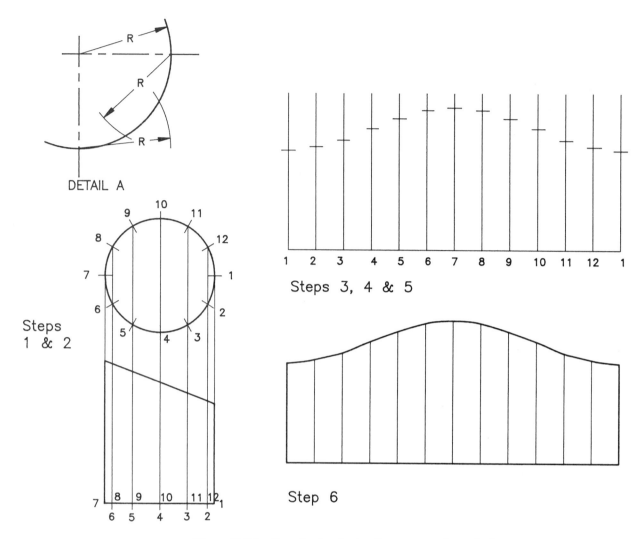

Figure 7-24 Steps in developing flat pattern of a cylinder.

3. Draw a stretch-out line with a length equal to the circumference, or 2 * R * 3.14. Divide this line into 12 equal spaces by the method presented in Section 3-2, page 78 or by using dividers to transfer distances from the top view.

4. From each point draw a construction line perpendicular to the stretch-out line.

5. Transfer the length of each projection line, as shown in the front view, to the development.

6. With your irregular curve connect the upper points. To complete the development, darken the end lines and stretch-out line and erase unnecessary marks.

The techniques for drawing developments of pyramids, cones, transition pieces, and other shapes are left to more advanced texts.

7-7 CONCLUSION

This chapter has described techniques of drawing and clarifying complex objects. It has covered auxiliary views, section views, removed views, and developments. However, you have had only a brief look at each of these. If you take more advanced courses in technical drawing, you will learn much more about these techniques. In fact, advanced study in technical drawing is, in large part, the study of specialized methods for solving graphical problems and depicting unusual objects.

In case you haven't already figured it out, the flat pattern in the opening illustration is for a sheet metal bookend.

Check your general understanding of the topics covered in this chapter by answering the following questions.

REVIEW QUESTIONS

1. For a surface to appear true size, it must be _____ to the observer's line of sight.

   a. parallel

   b. inclined

   c. perpendicular

   d. oblique

2. For a surface to appear as an edge, it must be _____ to the observer's line of sight.

   a. parallel

   b. inclined

   c. perpendicular

   d. oblique

3. An auxiliary view is different from a standard view because:

   a. none of the below

   b. it shows none of the object's dimensions

   c. it shows one principal dimension instead of two

   d. it shows two principal dimensions instead of three

4. Primary auxiliary views are used to show _____ surfaces TS&S.

    a. normal

    b. inclined

    c. oblique

    d. curved

5. Secondary auxiliary views are used to show _____ surfaces TS&S.

    a. normal

    b. inclined

    c. oblique

    d. curved

6. Crosshatching is usually drawn _____ apart and at an angle of _____.

    a. 3 mm; 30°

    b. 10 mm; 45°

    c. 10 mm; 30°

    d. 3 mm; 45°

7. Section views:

    a. may be shown anywhere on a drawing

    b. seldom show hidden lines

    c. show ribs and spokes with no crosshatching

    d. all the above

8. Which type of section view shows part of an object in section and part as a standard view?

    a. full

    b. offset

    c. half

    d. aligned

9. Which type of section view is shown within the part outline of a standard view?

    a. offset

    b. full

    c. aligned

    d. revolved

10. Which type of section view has a cutting plane that is bent at right angles more than once?

    a. full

    b. offset

    c. aligned

    d. revolved

11. Which type of section view is not a true projection—that is, its features are moved as needed?

    a. full

    b. half

    c. aligned

    d. revolved

12. Which of the following geometric forms would be impossible to draw a flat pattern for?

    a. prism

    b. pyramid

    c. cone

    d. sphere

PROBLEMS

7-1 Match each multiview with the correct auxiliary view.

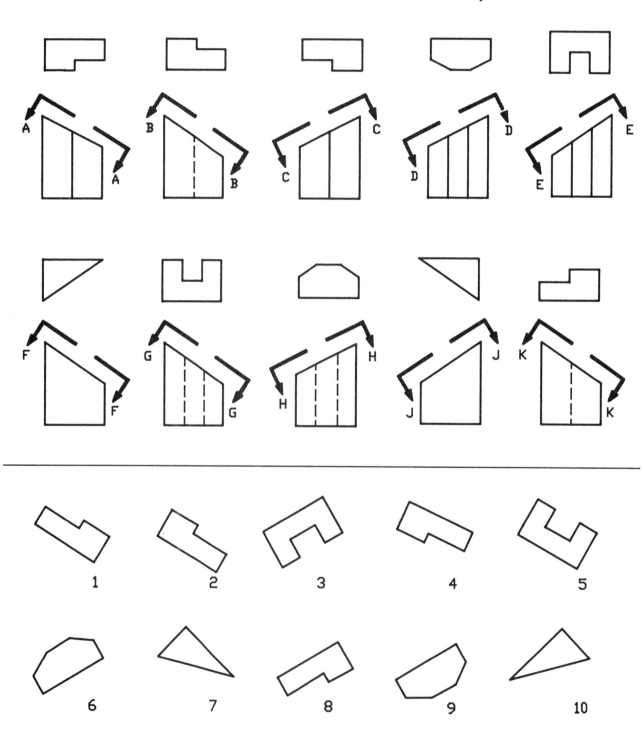

7-2   Select the auxiliary views possible from the direction of sight indicated by the arrow in the given view.

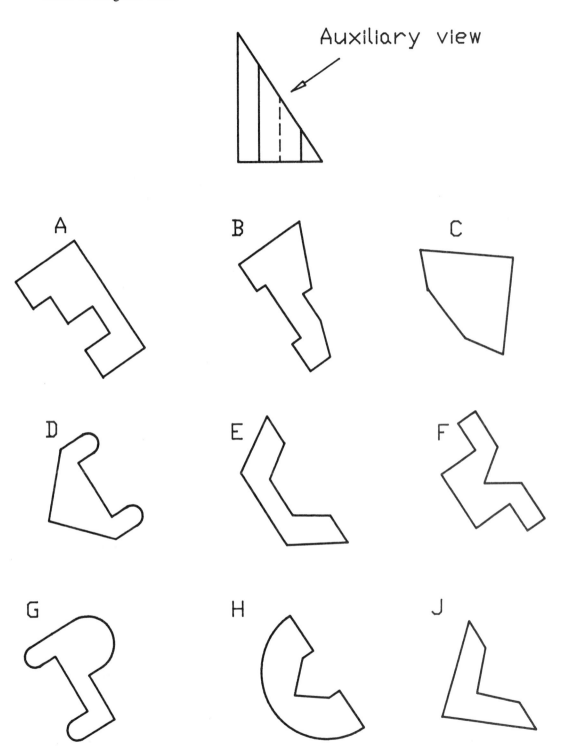

Auxiliary view

A

B

C

D

E

F

G

H

J

7-3 Transfer the given views to graph paper. Draw auxiliary views for each multiview, as indicated by the projection lines.

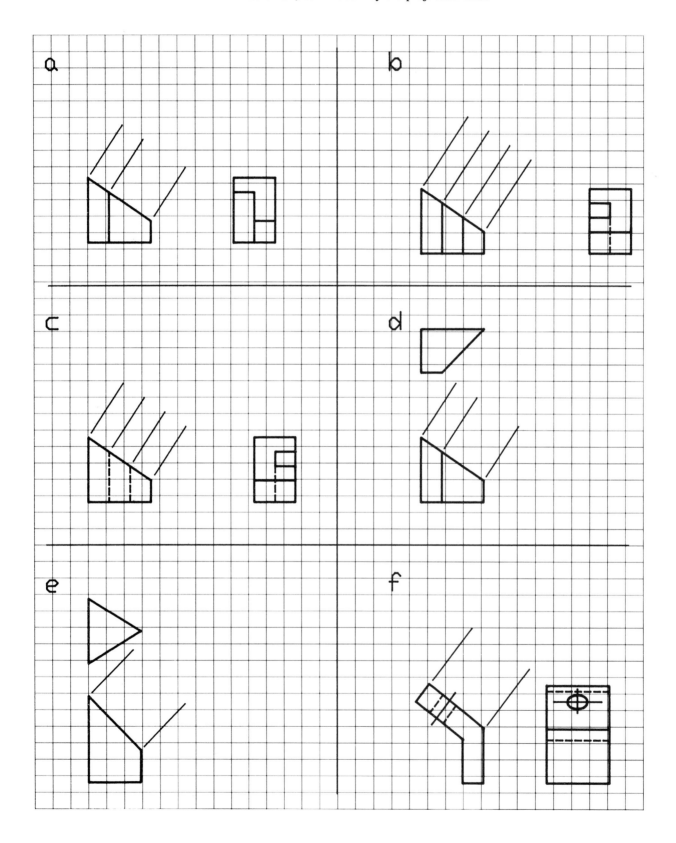

7-4 Transfer the given views to graph paper. Draw auxiliary views of the inclined
surfaces.

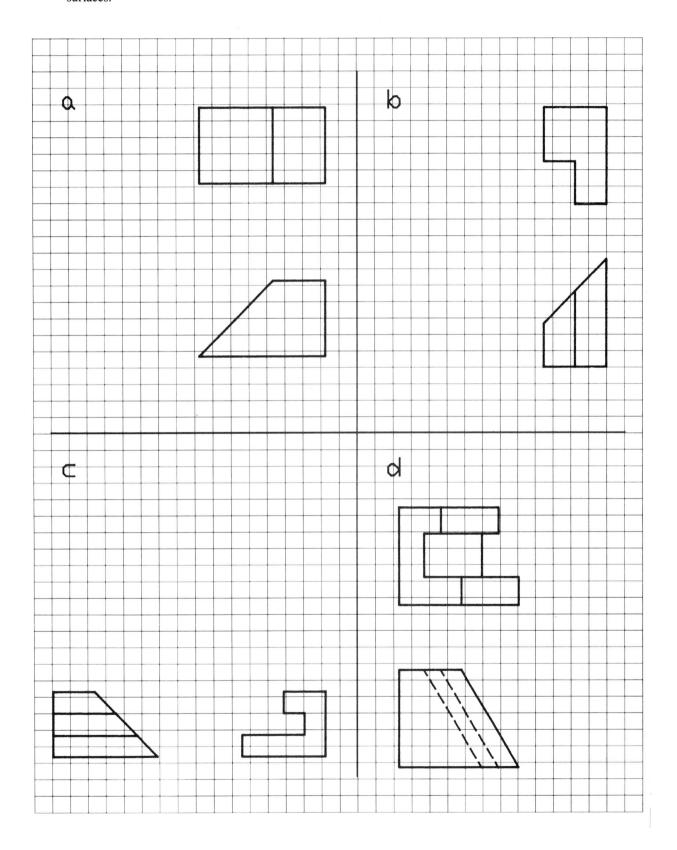

7-5 Redraw the front view, replace the top view with an auxiliary view that shows the inclined surface true size and add a right side view.

a

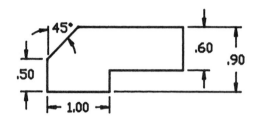

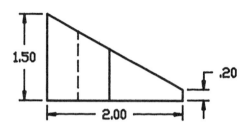

b

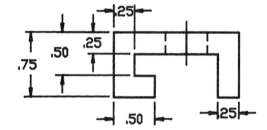

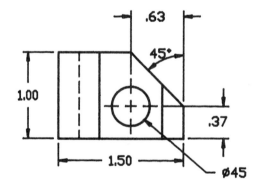

7-6 Match each section view with the correct multiview.

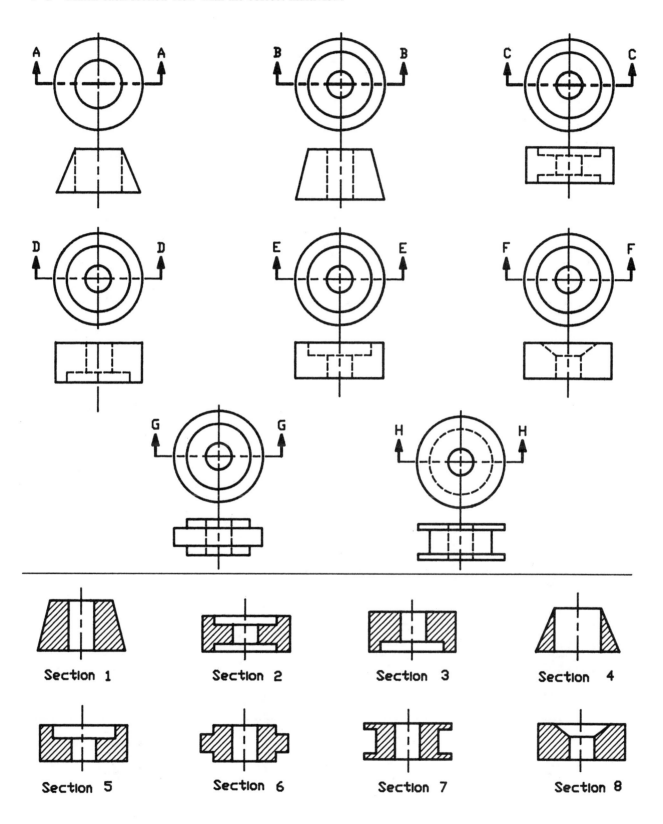

7-7 Match each section view with the correct multiview.

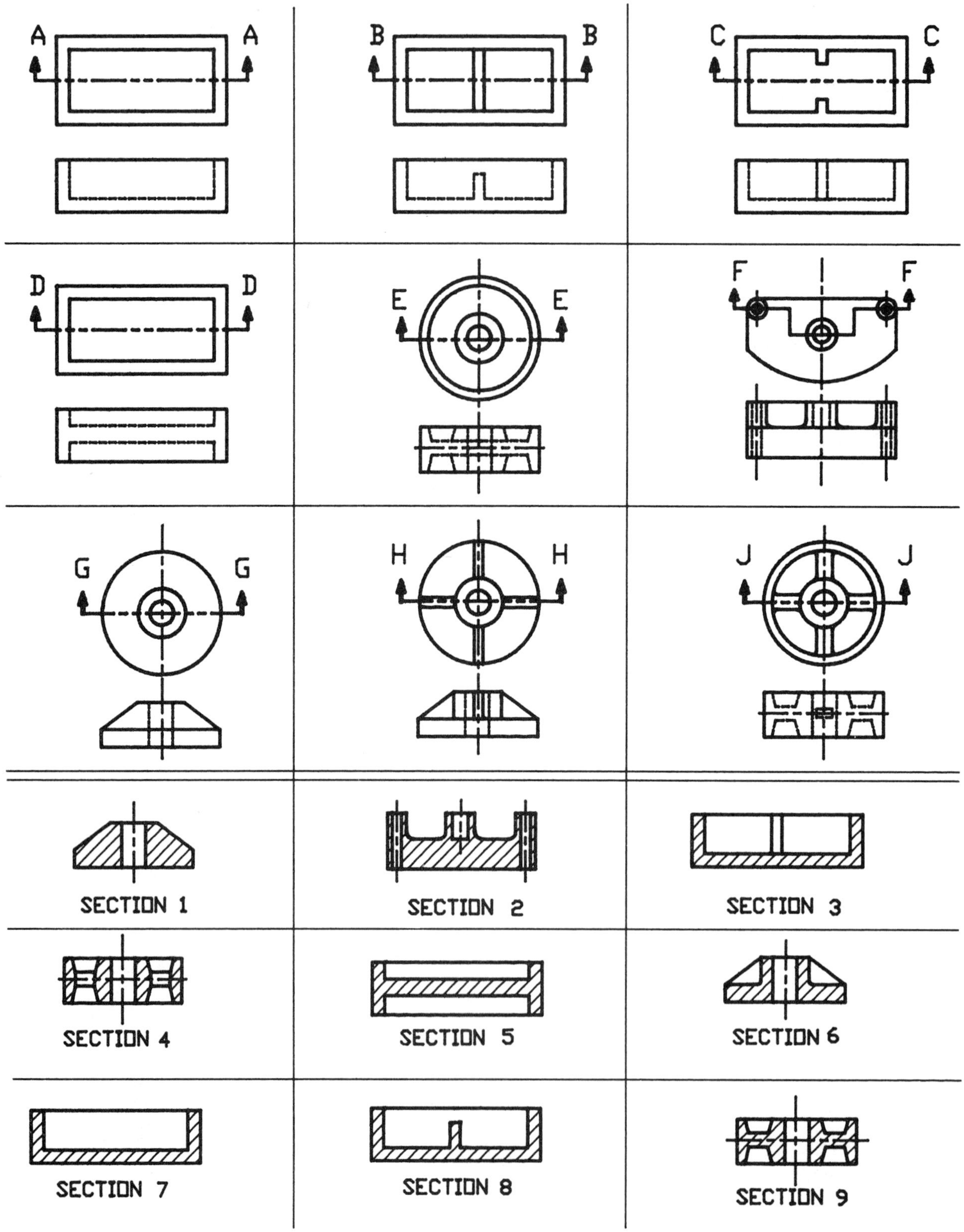

SECTION 1

SECTION 2

SECTION 3

SECTION 4

SECTION 5

SECTION 6

SECTION 7

SECTION 8

SECTION 9

7-8 Select the section views possible from the cutting plane shown in the given view.

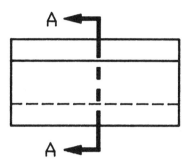

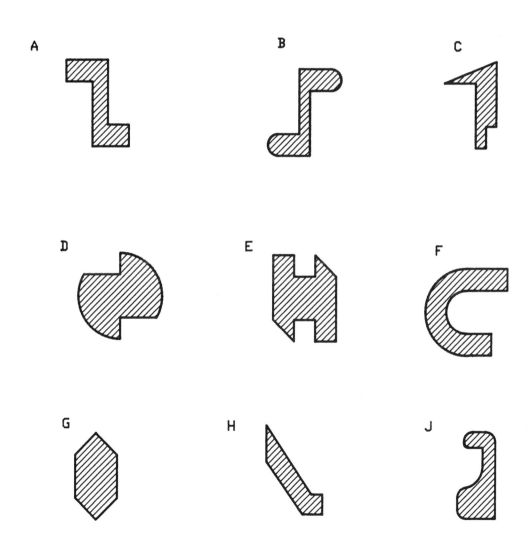

7-9 Transfer the given views to graph paper. Draw section views and label them as indicated.

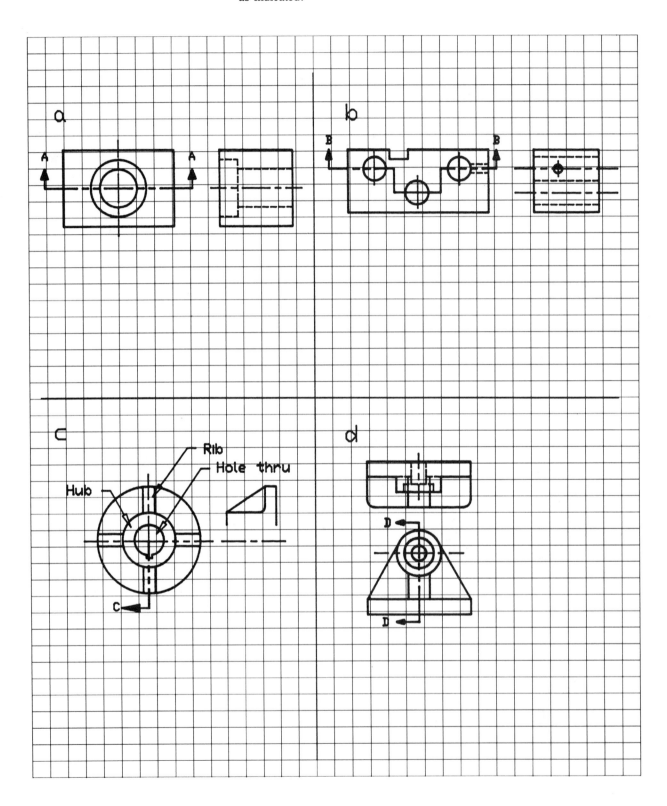

7-10   Transfer the given views to graph paper. Draw section views and label them
as indicated.

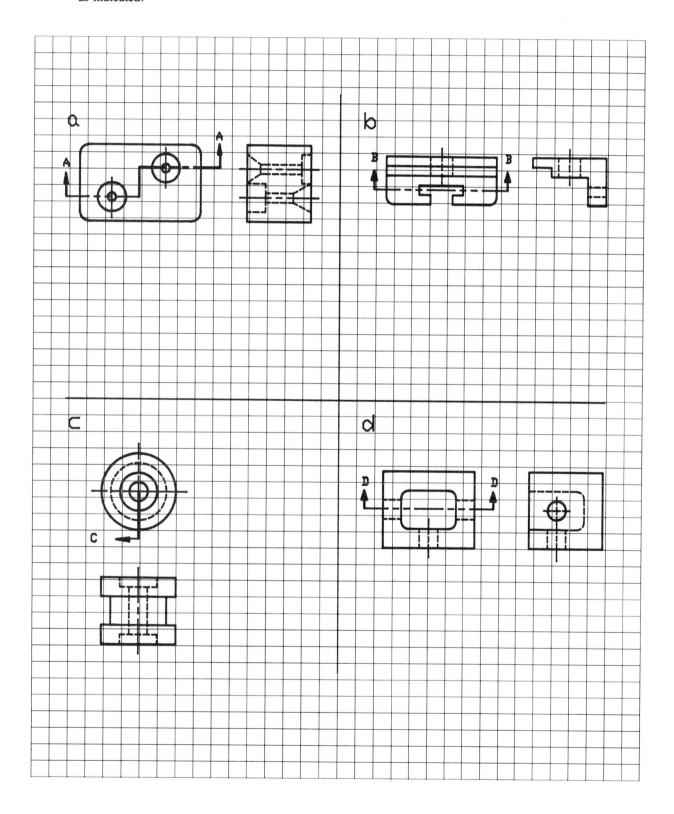

7-11   a. Draw top, front, and full section views as indicated by the arrows.

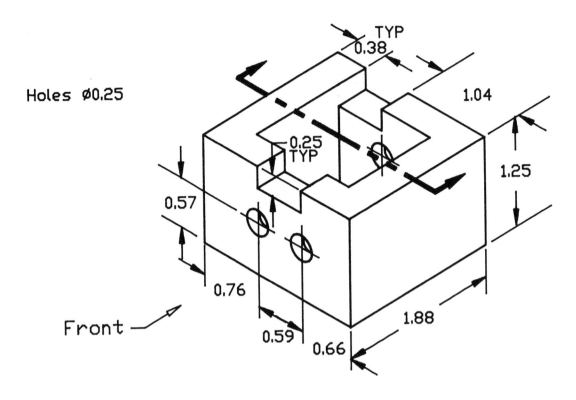

Holes ⌀0.25

b. Draw top and section views as indicated by the arrows.

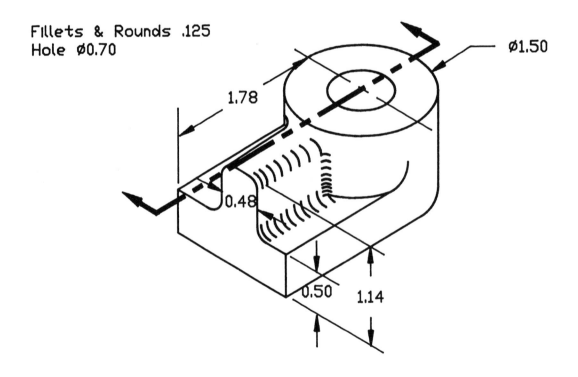

Fillets & Rounds .125
Hole ⌀0.70

7-12 Match each flat pattern with the correct multiview.

1

2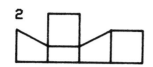

3

4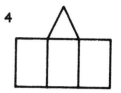

5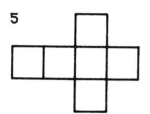

6

7

8

9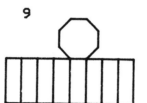

10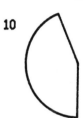

A   B   C   D   E

F   G   H   J   K

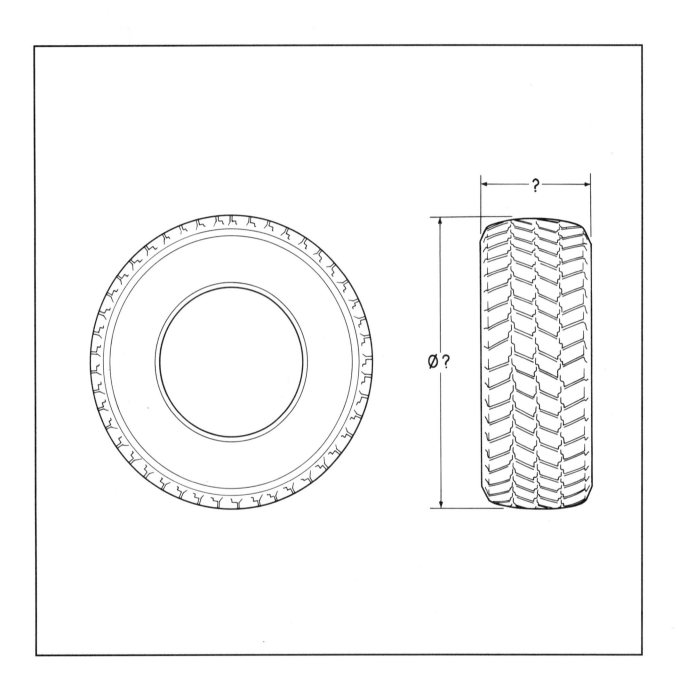

DIMENSIONS
AND SPECIFICATIONS

8-1 INTRODUCTION

The emphasis thus far in this book has been on shape description. It is very important in technical drawing to create the best possible description of an object's shape. But no matter how well you draw, shape description will not describe an object's size. For that you must use size description, or dimensioning. The illustration that opens this chapter shows an ordinary tire. Or does it? Can you tell if it is ordinary or not, without knowing its size? It could be an automobile tire with a diameter of 2 ft or an off-road truck tire with a diameter of 10 ft. Without numbers, it is impossible to tell. So the two parts of technical drawing, shape description and size description, are both needed to provide complete descriptions of objects.

This chapter covers size description. It presents the principles and techniques for applying dimensions and specifying needed information that cannot be communicated graphically. The information discussed includes dimensional units and quantities, dimensioning principles and techniques, thread specifications, principles of tolerancing, and surface texture specifications.

After you complete this chapter, you should:

1. know the commonly used linear measurement units

2. understand and be able to apply the three principles— completeness, usability, and readability—in dimensioning

3. know the correct techniques for using dimension and extension lines and leaders and for specifying quantities and units

4. know the accepted ways of dimensioning standard features

5. know the commonly used symbols and abbreviations

6. be able to specify common thread forms

7. understand the need for and methods used in specifying tolerances

8. know the basic symbols used in specifying surface texture

8-2 DIMENSIONAL UNITS AND QUANTITIES

Sizes and locations are shown on technical drawings by means of dimensions. **Dimensions** are the written information from which you can take and make measurements. Dimensions are composed of two parts: units and quantities. Units describe the type of measurement, either linear or angular, and quantities indicate the number of units in each dimension.

The linear units used most commonly in technical drawings are millimeters, meters, inches, and feet. They are indicated on each drawing in one of several ways. On some drawings the unit is indicated next to the dimension by means of abbreviations such as "mm" or "in." On most drawings, however, the units are indicated by means of one note specifying units for all the dimensions on that drawing. The individual dimensions show only the quantity. Some drawings are dual dimensioned, which means each dimension is indicated by two sets of units and quantities. For example, you may find drawings dimensioned in both millimeters and inches. This practice was once common but is presently discouraged.

The degree is the most common angular unit used. A degree is 1/360 of a complete circle. It may be subdivided into decimal parts, such as .10 or .50 of a degree, or into 60 parts called minutes, each of which may be divided into 60 smaller parts called seconds.

The linear and angular units used in a particular drawing depend on the type of drawing, as well as on its country of origin. Table 8-1 lists the normal units used in drawings for manufacturing and construction. Metric units (meters and millimeters) are used exclusively in nearly all parts of the world

Table 8-1 Commonly used dimensional units and quantities

Manufacturing drawings		
Units	Quantity examples	Notes
Millimeters	22 31.4 0.7	Decimal fractions used. No symbols used.
Inches	4.00 28.65 .225	Decimal fractions used. Inch marks not used.
Degrees	25.5°	Decimal fractions used. Degree symbols used.

Architectural and construction drawings		
Units	Quantity examples	Notes
Feet and inches	6'-2½" 7'-0"	Common fractions used. Symbols (' and ") used.
Feet and decimal Fractions of a foot	22.15' 124.30'	Used on plats, maps, and structural drawings. Symbol (') used.
Meters	15.72 31.5	Not commonly used in the United States.
Degrees	95°25'30"	Minutes and seconds used. Symbols (°, ', and ") used.

except the United States. Here, feet and inches, as well as metric units, are used.

The quantity parts of dimensions seldom work out to be whole numbers; usually they have fractional parts. These are stated in decimals, except on architectural drawings, where feet, inches, and fractions of inches are used. The inch is fractionally divided on some older manufacturing drawings, also. Table 8-1 illustrates commonly used units and quantities.

8-3 PRINCIPLES OF DIMENSIONING

How objects are dimensioned is dependent on several factors. Among these are the object's geometry, its function, and the precision requirements. You must keep these factors in mind when applying dimensions. It is best to spend some time analyzing the object and planning the dimensioning scheme before you start the work. Since objects can have an almost infinite number of geometries and functions, it is difficult to lay down precise rules for dimensioning all objects. Therefore, to achieve proper dimensioning, you must apply some principles and general techniques and use good judgment. Seldom is one method the only right one for dimensioning a particular object. Rather, several methods are probably right, but only one is the best method. The best method can nearly always be determined by considering the following principles.

Completeness. Objects must be dimensioned completely, without duplicating any dimensions. All the sizes and features required for making the object must be shown on the drawing, but no features should be dimensioned more than once. See Figure 8-1.

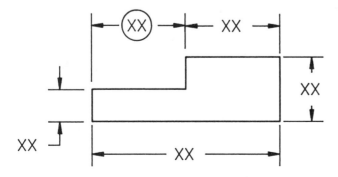

Figure 8-1 This object is completely dimensioned, but the circled dimension is a duplication and should be eliminated.

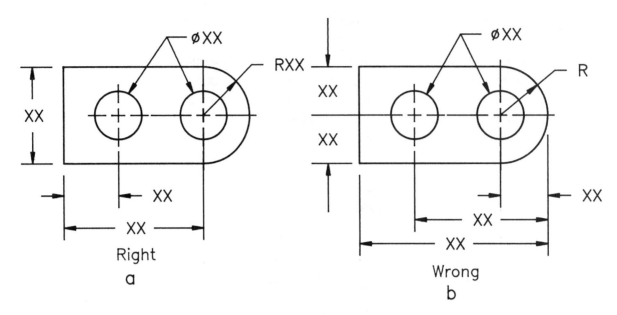

Figure 8-2 The drawing in part (a) is more usable than the one in part (b) because, in (a), dimensions refer to features that are easy to measure from and calculations are minimized.

Usability. The dimensions cited must be the most useful for producing the object. Show dimensions that relate to features that are easy to measure from and that minimize calculations by users of the drawing. See Figure 8-2.

Readability. Dimensions should be shown in the contour view or the view that best describes the feature or shape being dimensioned. Place dimensions as close to the associated features as possible. Group dimensions between views, if possible. Don't measure from hidden lines. See Figure 8-3.

Figure 8-3 Compare the right and wrong versions of dimensions (a), (b), and (c). The right dimensions are more readable because they are placed in the views that best show the associated features.

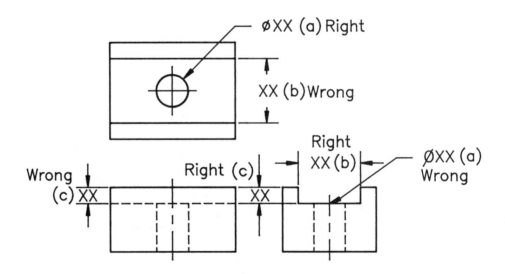

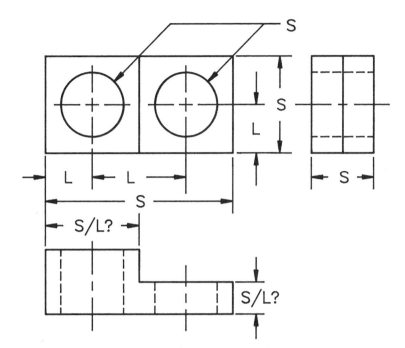

Figure 8-4 Size dimensions are indicated by "S" and location dimensions by "L." Dimensions that might indicate either size or location are indicated by "S/L?" Note the grouping of the dimensions and their placement between the views.

In planning the dimensioning of a drawing, it is often useful to categorize dimensions as either size or location dimensions. Size dimensions tell the size of features — that is, size dimensions tell length, width, diameter, or the like. Location dimensions tell where features — holes, slots, notches and so on — are located relative to the ends, edges, or other features. Figure 8-4 shows examples of size and location dimensions. These distinctions are helpful in the planning of dimensioning because they help ensure that the dimensioning will be complete. Thinking in terms of size and location gives you a systematic way of approaching the task.

8-4 GENERAL DIMENSIONING TECHNIQUES

To attain maximum clarity in dimensioning, you must use approved techniques. Dimensioning practices in the United States are governed by the American National Standards Institute (ANSI). This organization maintains the majority of standards used by American industry. The document that contains dimensioning standards is ANSI Y14.5M–1982. Figure 8-5 shows an example of a dimensioned drawing that conforms to ANSI Y14.5M–1982. The dimensioning standards define the correct methods for showing the extent, direction, and location of dimensions (see Figure 2-29, page 51) and the correct placement and sizing of numbers and letters.

Figure 8-5 A drawing dimensioned in accordance with ANSI Y14.5—1982. Note the use of the diameter symbol (∅) and the abbreviation "R" for *radius*. The label "2 × " indicates that the dimension occurs in two different places in the drawing.

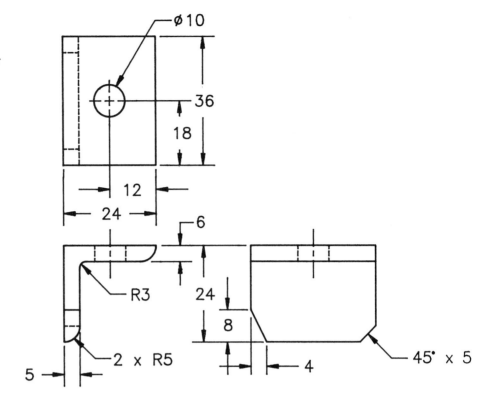

Extension Lines

Extension lines (sometimes called projection lines) are thin lines used to extend edges, surfaces, or points outside an outline. They are used because it can be confusing when dimensions are applied directly to object lines or inside a part outline. Extension lines normally have a small gap (1.5 mm) between them and the part outline, and they extend at least 3 mm beyond the farthest related dimension line. They should be placed to minimize length and line crossovers. When crossovers are unavoidable, do not break the extension lines. Figure 8-6 shows the proper application of extension lines.

Dimension Lines

Dimension lines are thin lines, usually with small arrowheads at each end, that indicate the direction and extent of dimensions. They are always parallel to the direction of the dimension and perpendicular to the associated extension lines. The arrows should be small and proportioned as shown in Figure 8-7, and they should just touch their extension lines. On drawings for manufacturing, dimension lines are broken so that the numerals may be written in the resulting space. Dimension lines in construction drawings do not always end with arrows. They may be unbroken and the numerals may be

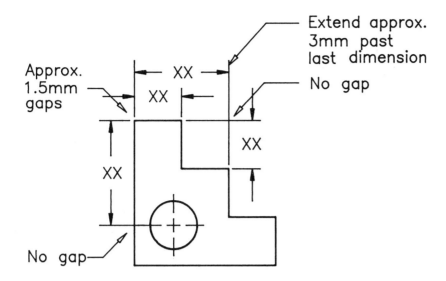

Figure 8-6 Examples of properly drawn extension lines.

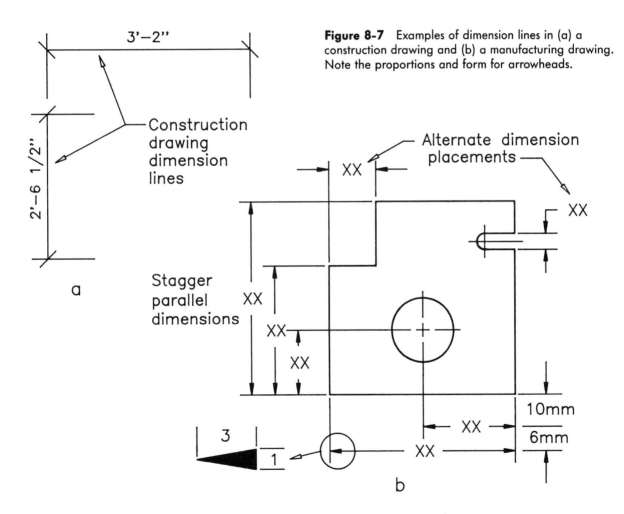

Figure 8-7 Examples of dimension lines in (a) a construction drawing and (b) a manufacturing drawing. Note the proportions and form for arrowheads.

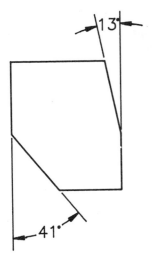

Figure 8-8 Examples of angular dimensions.

written above or beside the line. Dimension lines should be placed and spaced as shown in Figure 8-7. If a series of parallel dimensions is shown, the breaks should be staggered. Notice that the shortest dimensions are placed closest to the object. If the dimensioned space is small, dimension lines and numerals may be placed outside the extension lines by any of the methods shown in Figure 8-7. Dimension lines should never cross.

Dimension lines for angles are arcs that are drawn using the vertex of the angle as the center (see Figure 8-8).

Leaders

Leaders, or leader lines, are thin lines with an arrow at one end. Leaders are used to connect a note or dimension with a feature of an object. The leader line is really two connected lines: One is short and horizontal and the other is slanted, with an arrow at its end pointing to the feature. The short horizontal shoulder leads to the note or dimension (see Figure 8-9). If a leader points to a circle or arc, it should be drawn radially, pointing at the center, with the arrow just touching the circle or arc.

Dimension Numerals and Notes

All numerals and notes used on technical drawings should be horizontal—that is, parallel to the top and bottom of the drawing sheet. This is called unidirectional dimensioning. The other method, aligned dimensioning, where the numerals line up with their dimension lines, is seldom used except on con-

Figure 8-9 Examples of the forms and uses for leaders.

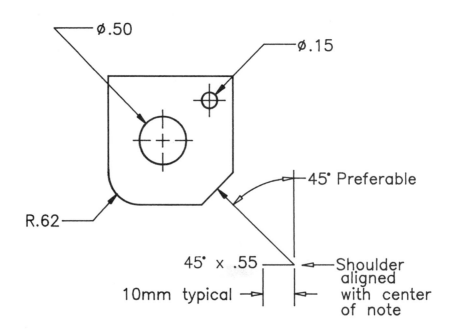

struction drawings. All numerals must show the appropriate form for the type of units represented. For millimeters this consists of no decimal point for whole numbers and a zero shown to the left of the decimal point when the dimension is less than one. For decimal inch dimensions the appropriate form consists of a minimum of two decimal places on all dimensions and no zero to the left of the decimal point for dimensions less than one. See Table 8-1 and Figure 8-10 for examples. Note that architectural dimensions contain feet and inches and common fractions of inches, as shown in Table 8-1.

The size of the numerals varies according to the size of the drawing sheet. For A3, A4, A, and B sheets, the lettering should be 3.5 mm or .125 in. For all larger sheets, it should be 5 mm or .15 in. (see Table 1-1, page 16).

8-5 SYMBOLS AND ABBREVIATIONS

Current dimensioning practice makes use of many symbols and abbreviations. These symbols and abbreviations are used to shorten and standardize notes, dimensions, and instructions commonly placed on technical drawings. Table 8-2 lists the most commonly used symbols and abbreviations. Examples of their application are shown in the various figures in this chapter.

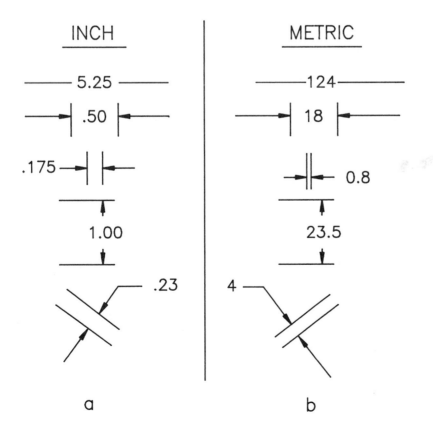

Figure 8-10 Appropriate form for (a) dimensioning in inches and (b) dimensioning in metric units.

Table 8-2   Dimensioning symbols and abbreviations

Current	Old	Meaning
∅	D or Dia	Diameter
S∅	SD or S Dia	Spherical diameter
R	R	Radius
CR	None	Common radius
SR	SR	Spherical radius
$\overset{\frown}{20}$	None	Arc dimension
2×	2 Places	Number of times/places
(34)	34 Ref	Reference dimension
45̰	45̰ or NTS	Dimension not to scale
⊔	Cbore/SF	Counterbore/spotface
∨	CSK	Countersink
⊤̄	Deep	Depth/deep
□	SQ	Square shape
✓	✓	Surface texture

8-6 DIMENSIONING FEATURES

Many features—such as holes, slots, keyways, and chamfers—are so common in the geometry of most objects that standard methods of dimensioning them have been developed. The following sections describe these standard techniques for the most commonly encountered features.

Chamfers

Chamfers occur when 90° corners are cut off at some other angle. They may be dimensioned by using either two linear dimensions, as shown in Figure 8-11a, or an angle and a linear dimension, as shown in Figure 8-11b. Two lin-

Figure 8-11   Three methods for dimensioning chamfers.

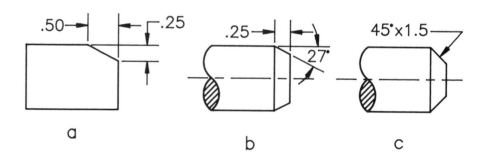

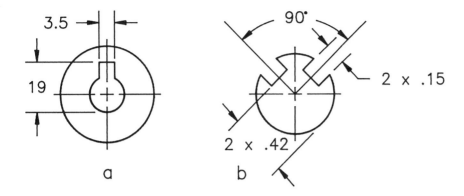

Figure 8-12 Dimensioning a keyway (a) in a hole and (b) in a shaft.

ear dimensions are preferred when close control of the chamfer size is necessary. When the angle of the chamfer is 45°, it may be dimensioned by a note, as shown in Figure 8-11c.

Keyways

Keyways (also called keyseats) are notches cut in a part for the purpose of locating or fastening other parts in an assembly. Keyways require at least three dimensions: width, depth, and location (see Figure 8-12). If the keyway is on a shaft or in a hole, then the depth is dimensioned from the opposite side of the shaft or hole.

Circles and Arcs

Circles are always dimensioned by their diameters (the diameter symbol, Ø, precedes the numbers), and arcs are always dimensioned by their radii (the abbreviation R precedes the numbers). Both circles and arcs are normally dimensioned in the profile view. But, if there are several concentric circles on an object, they should be dimensioned in a side or longitudinal view, as shown in Figure 8-13a. Otherwise, circles should be dimensioned in a view that shows them as circles, as shown in Figure 8-13, parts b and c. Notice in Figure 8-13b that the leader is radial (points to the center of the circle) and that the arrow just touches the circle.

Arcs can be dimensioned by either of the methods shown in Figure 8-14a if the arc is defined by some means other than the center (such as a tangency point). If the center is needed, it must be defined by linear dimensions and the radius leader should pass through the center, as shown in Figure 8-14b. Figure 8-14c shows a spherical radius callout.

Figure 8-13 (a) Dimensioning concentric circles in a side view. (b) and (c) Dimensioning circles in views that show them as circles.

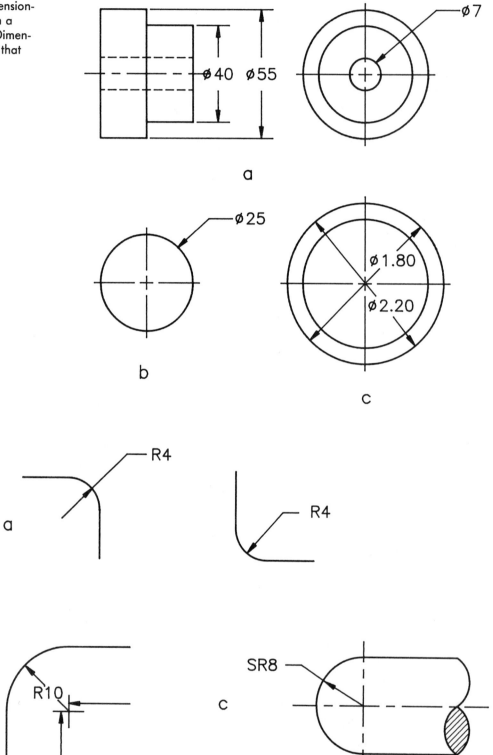

Figure 8-13: a, b, c

Figure 8-14 (a) Dimensioning an arc without using the center. (b) Dimensioning an arc using the center. (c) Using a spherical radius callout to dimension an arc.

Round Holes

Round holes are dimensioned by their diameters and are assumed to go through the object unless a note is added indicating otherwise. A hole that does not go through is called a **blind hole,** and the depth dimension is the depth of the full diameter measured from the surface of the part.

Round holes may be modified by counterboring, countersinking, and spotfacing. In all three cases the dimensions are normally given by callout in the profile view, as shown in Figure 8-15 (see also Table 8-2). However, it is permissible to show depth dimensions in a side view, if necessary. Counterbored holes require diameters for the holes and counterbores and a counterbore depth. Countersunk holes require diameters for the holes and countersinks and an included angle for the countersink. Spot-faced holes require diameters for the holes and spotfaces, but no depth dimension.

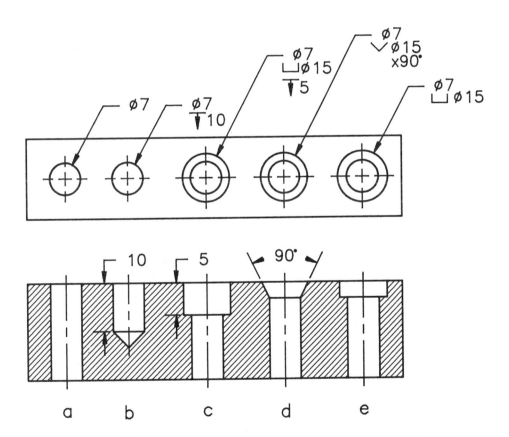

Figure 8-15 Dimensioning round holes: (a) a thru hole, (b) a 10-mm-deep blind hole, (c) a counterbored hole, (d) a countersunk hole, and (e) a spot-faced hole.

Figure 8-16 Two methods for dimensioning a slotted hole.

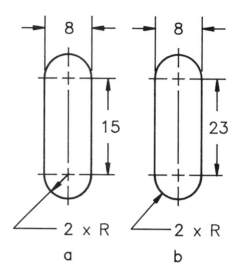

Slotted Holes

Slotted holes may be dimensioned by either of the methods shown in Figure 8-16. Notice that the callout R is used to indicate that the ends of the slot are radiused, but no value is given. Slots are dimensioned this way to avoid duplicating the width dimension.

Repetitive Features

Features that occur more than once on an object (repetitive features) require special dimensioning techniques to prevent confusion on the drawing and to minimize the amount of drafting that must be done. There are two fundamental methods for defining the placement of features: using rectangular coordinates and using polar coordinates. **Rectangular coordinates** are used for objects with feature-placement patterns that are rectangular—that is, they require linear dimensions only. **Polar coordinates** are used with circular or radial objects and with patterns that require angular dimensions. Figure 8-17 shows examples of rectangular coordinates and Figure 8-18 shows polar coordinates. The rectangular method provides better dimensional control, but the polar pattern is often easier to draw.

Repetitive features of the same size are dimensioned only once; a note indicates how many places the feature occurs, or the number of places it occurs is included as part of the dimension. (The number of places is followed by a multiplication symbol. The note "3×" is read as "three times," for example.) The same general method is used to provide location dimensions, as Figure 8-18 shows. These techniques reduce the number of dimensions, without reducing the clarity of the drawing.

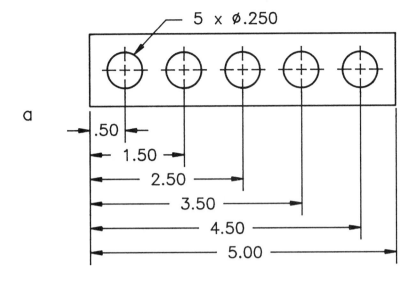

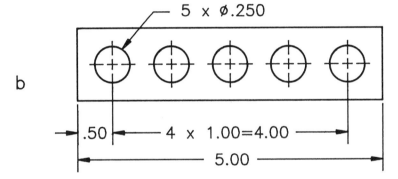

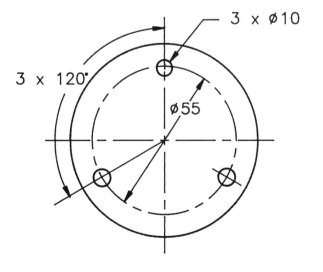

Figure 8-17 Dimensioning repetitive features by using rectangular coordinates. The drawing in (a) is more detailed than the drawing in (b), but drawing (a) requires more effort and space than drawing (b).

Figure 8-18 Polar coordinate dimensioning of repetitive features.

8-7 THREAD SPECIFICATIONS

Threads are a common type of geometry found on cylindrical parts and in round holes, yet the complex shape of a thread makes it very difficult to draw. Fortunately, schematic and simplified representations and dimensional and geometric callouts are used, and these eliminate the need to make anything approaching a detailed drawing of a thread.

Figure 8-19 illustrates the basic aspects of thread geometry—aspects that are common to various styles and sizes of threads. Figure 8-20 shows profiles of some of the most commonly used thread forms. Each of these has a specific purpose. In this limited coverage of threads, the primary interest is in metric and unified forms, because they are the most widely used.

As stated earlier, threads have very complex geometry. The basic form is called a helix. Threads are either external (on the outside of a cylinder) or internal (on the inside of a hole). They may be single (continuous thread), double (two separate parallel threads), or triple (three separate parallel threads), and they are either left-hand or right-hand turning. As shown in Fig-

Figure 8-19 Thread geometry.

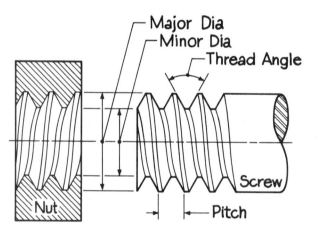

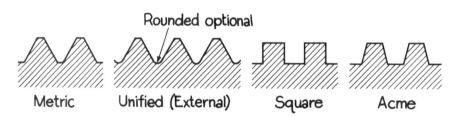

Figure 8-20 Four commonly used screw thread forms.

ure 8-20, their cross-sectional profile may take a variety of forms. They may be made to a variety of precision standards called **fits** and are available in a wide range of sizes. All this information, though difficult to show by drawing, is easy to communicate through the use of standardized callouts or notes.

Figure 8-21a shows the basic callout format for metric threads; Figure 8-21b shows the format for unified threads. The metric thread callout is obviously much simpler. The letter *M* always precedes the diameter, which is followed by a multiplication symbol (×) and the pitch. The **pitch** is the distance between any two adjacent threads. Unified thread callouts indicate threads per inch rather than pitch, and they distinguish between external and internal threads by appending an *A* (for external threads) or a *B* (for internal threads) to the fit number. Metric thread callouts include fit information for close-tolerance threads only. Both thread types use the abbreviation LH at the end of the callout to signify a left-hand thread. The lack of a designation indicates a right-hand thread.

a
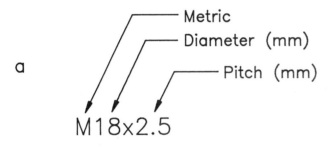

b
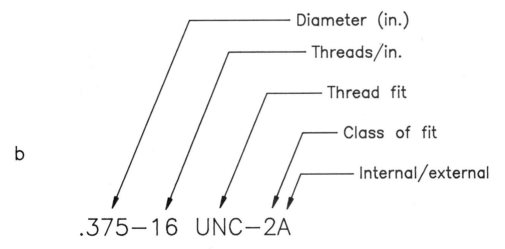

Figure 8-21 Screw thread callouts: (a) metric and (b) unified.

Actual detailed drawings of threads are complex and difficult to draw (see Figure 8-22a). Therefore, threads are usually drawn symbolically.

Threads are symbolized in two different ways: in schematic and simplified representations. Figure 8-22, parts b and c, shows both internal and external threads. The simplified drawing is preferred because it takes less work.

This chapter has covered only the most basic ideas about the most commonly used threads. For a complete discussion, see ANSI standards Y14.6–1978 and ANSI Y14.6aM–1981.

8-8 TOLERANCING

A **tolerance** is a permissible variation in the size of a part or the location of a feature. Tolerances are necessary because it is impossible to manufacture parts to an exact size or shape. This is because it is impossible to measure anything exactly, as you have probably discovered while drawing. Actually there is a second reason for using tolerances, and that is the need for interchangeability among parts. Manufacturers need some means of ensuring that

Figure 8-22 Thread representation methods: (a) detailed (rarely used), (b) schematic, and (c) simplified (preferred).

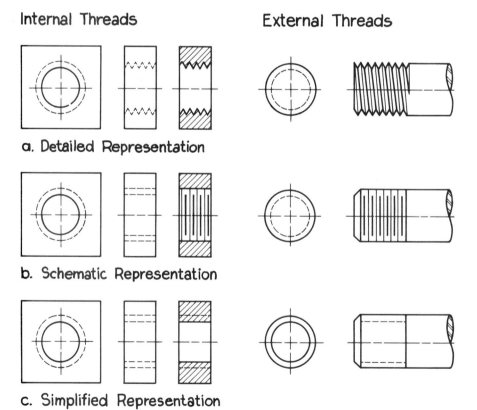

Internal Threads   External Threads

a. Detailed Representation

b. Schematic Representation

c. Simplified Representation

parts that are supposed to fit together (like nuts and bolts) *will* fit together. Properly chosen tolerances provide that means. This section will cover only the essentials of tolerancing. If you wish more information, refer to ANSI standard Y14.5M–1982.

Definitions

This discussion of tolerancing begins with a few definitions. **Basic size** refers to the theoretically exact size of a dimension. It is the size an object would be if it could be made exactly. It is also the size usually referred to when talking about a dimension. As you know, the tolerance is the total amount that the basic size is allowed to vary. **Limits** are the maximum and minimum sizes. (Another way of defining tolerance is as the difference between the limits.) **Datums** are theoretical points, lines, or planes, that are usually on the object and are used as starting positions for dimensions.

Specification of Tolerances

Tolerances may be specified in a variety of ways. They may be expressed as limits, as a basic dimension together with limits, as a basic dimension with plus (+) or minus (−) some portion of the tolerance, as a minimum (MIN) or maximum (MAX) dimension, or as an overall drawing tolerance. These methods are illustrated in Figure 8-23. Specifying an overall drawing tolerance is the most common method used because it results in the least amount of work and the cleanest drawing. In Figure 8-23e the notation ".XX" refers to the number of decimal places included in any dimension, and it is used as a code to indicate which tolerance applies to a particular dimension.

A dimension with limits specified both above and below the basic size is called bilateral tolerancing. However, the tolerance can be applied in one

a —— .503 ——
     .497

b —— .500 .503 ——
          .497

c —— .500 +/−.03 ——

d —— .503 MAX ——

e ALL TOLERANCES .XX +/− .01
                 .XXX +/− .005

Figure 8-23 Methods for specifying tolerances: stating (a) limits, (b) basic size and limits, (c) basic size and plus or minus some portion of the tolerance, (d) maximum/minimum dimension, and (e) overall drawing tolerance.

direction only, in which case the dimension can vary in one direction only—it can be either larger or smaller than the basic size, but not both. This is called unilateral tolerancing. In any case, the maximum limit is always written above or in front of the minimum limit. The use of MAX (or MIN) simply means that the dimension is not allowed to be any larger (or smaller) than the basic size, but it can vary any amount in the opposite direction.

Applying Tolerances

A tolerance applied to a dimension means that it is permissible for that dimension to be as big as the maximum limit, as small as the minimum limit, or anywhere in between. Any dimension outside the limits is said to be **out of tolerance** and is unacceptable. This is a simple concept as long as it is limited to single, independent dimensions. But it becomes very complex when applied to multiple, dependent dimensions.

The part shown in Figure 8-24a contains three holes, each 30 mm apart (¢ to ¢) and 30 mm from the ends (¢ to edge). The overall length is 120 mm. If a +/− 1-mm tolerance is applied and the dimensions all go to the maximum size, the part will have an overall length of 4 * (30 + 1), or 124 mm. But the maximum overall length can be only 120 + 1, or 121 mm! Is the part OK or should it be rejected? The answer is uncertain. This type

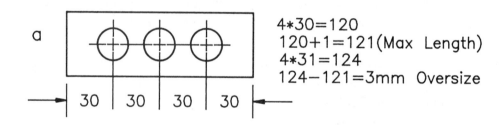

```
4*30=120
120+1=121(Max Length)
4*31=124
124−121=3mm Oversize
```

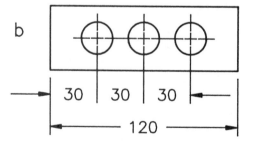

Figure 8-24 Examples of chain dimensioning.

of dimensioning is called chain dimensioning. Each hole location dimension is dependent on the previous dimensions for its actual location. In other words, the dimensions form a sort of chain. In chain dimensioning, the tolerances can accumulate, if they all go in the same direction, and result in a tolerance buildup (or tolerance stack-up) that probably exceeds the intended tolerance limits. Of course, you have probably noticed that Figure 8-24a is an example of poor dimensioning practice in that it gives no overall dimension. But even if corrected as shown in Figure 8-24b, you can see that there is still a problem. There can still be 93 mm from the end to the third hole, which is 2 mm more than is probably permissible.

Figure 8-25 shows a way out of this dilemma. Notice that all dimensions are independent of each other. A dimensional increase or decrease in any one does not affect any other. Each dimension starts at the right end, which has been selected as the datum, or baseline, for this set of dimensions. This style of dimensioning is called baseline dimensioning, and it should be used whenever there is a potential for tolerance buildups.

8-9 SURFACE TEXTURE

Surface texture is the condition of manufactured surfaces in terms of roughness and pattern. All production processes automatically create surfaces within predictable roughness ranges. If these are acceptable, no additional specification is required. For example, drilling holes always produces fairly smooth surfaces inside the holes. When this surface is acceptable, a level of texture is not normally specified on the drawing. Tolerances are very closely tied to surface texture. Very close tolerances automatically specify very smooth surfaces, so no special texture specification is required. It is only in special circumstances that surface texture must be indicated.

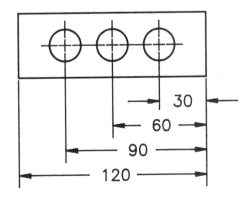

Figure 8-25 Figure 8-24 rendered using baseline dimensioning.

Figure 8-26 Surface texture symbols: (a) and (b) old-style symbols that are seldom used today. "M" means mill and "G" means grind. (c) Current symbols.

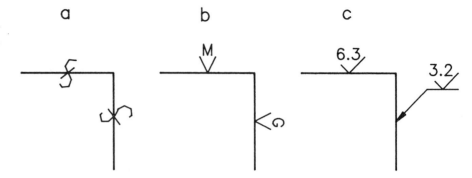

Finish marks are used to indicate special surface texture requirements. Over the years a number of different marks have been used; these are illustrated in Figure 8-26. Although they differ significantly in form, it is important to note that all marks are applied to edge views of the surfaces to be finished. The current version of the finish mark, shown in Figure 8-26c, can be used to indicate a variety of surface requirements, including roughness, waviness, and lay.

The most important attribute of texture is roughness height, which is defined as the average difference between the microscopic peaks and valleys that exist on any manufactured surface. This distance is measured in micrometers (μm) or microinches (μin.) (millionths of a meter and millionths of an inch). The range of standard specified surface roughness values is shown in Table 8-3.

Table 8-3   Standard specified surface roughness values

Roughness			
μm	μin.	Kind of surface	Usage
12.5	500	Rough	Used where vibration or stress concentration are not critical and close tolerances are not required
6.3	250	Medium	For general use where stress requirements and appearance are of minimal importance
3.2	125	Average smooth	For mating, with bolts or rivets, surfaces of parts with no motion between them
1.6	63	Smoother-than-average finish	For close fits or stressed parts except rotating shafts, axles, and parts subject to vibration
0.8	32	Fine finish	Used for applications such as bearings
0.4	16	Very fine finish	Used where smoothness is of primary importance, such as with high-speed shaft bearings
0.2	8	Extremely fine finish	Used for parts such as surfaces of cylinders in engines
0.1	4	Superfine finish	Used on areas where surfaces slide and lubrication is undependable

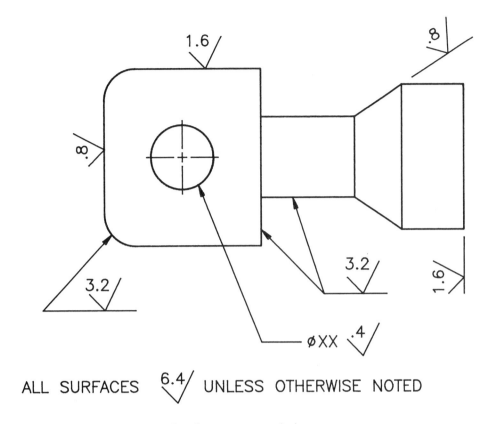

ALL SURFACES 6.4/ UNLESS OTHERWISE NOTED

Figure 8-27 Applications of surface texture symbols.

Figure 8-27 shows some typical applications of finish marks. For more information on this topic, see ANSI Y14.36−1978.

8-10 CONCLUSION

This chapter has taken a brief look at various aspects of size description. It has covered dimensional units and quantities, dimensioning principles and practices, thread callouts, the principles of tolerancing, and surface texture callouts—a very complex group of subjects. In a book of this type, it is possible to introduce only the fundamentals of size description. For a complete treatment of these most important subjects, you will need to study the ANSI standards manuals or more advanced texts.

Check your general understanding of this chapter by answering the following questions.

REVIEW QUESTIONS

1. The units commonly used in manufacturing drawings are

    a. feet and inches

    b. meters and millimeters

    c. feet and meters

    d. inches and millimeters

2. What dimension should be applied to a 200-mm length that is drawn one-half scale?

    a. 100 mm

    b. 200 mm

    c. 400 mm

    d. none of the above

3. Which of the following is *not* one of the three principles of dimensioning?

    a. functionability

    b. usability

    c. completeness

    d. readability

4. Which of the following are true statements concerning dimension and extension lines?

    a. Extension lines may cross each other if crossing is unavoidable.

    b. At their ends dimension lines have arrows that touch extension lines.

    c. Extension lines always have a gap between them and their associated object lines.

    d. All the above.

5. The presently accepted method of showing dimensions and notes is called:

    a. aligned dimensioning   construction

    b. unidirectional dimensioning

    c. bidirectional dimensioning

    d. misaligned dimensioning

6. Radii are always used in dimensioning:

    a. circles

    b. arcs

    c. cylinders

    d. holes

7. It is *not* good dimensioning practice to:

   a. dimension to hidden lines

   b. overlap dimension and extension lines

   c. place dimensions inside the part outline

   d. all the above

8. Which of the following are the currently used symbols for diameter counterbore?

   a. ⌀ ∨

   b. D-C bore

   c. ⌀ ⊔

   d. DIA CBORE

9. The callout M10×2 indicates a:

   a. machined surface finish

   b. medium-pitch unified thread

   c. modified countersunk hole

   d. metric thread

10. The total amount that a dimension may vary is called the:

    a. variance

    b. limits

    c. tolerance

    d. allowance

11. Accumulation of tolerances will *not* occur with _____ dimensioning.

    a. baseline

    b. chain

    c. continuous

    d. unilateral

12. A 250 μin. surface finish is:

    a. a very smooth finish

    b. equivalent to a 1.6 μm finish

    c. a medium finish

    d. equivalent to a 12.5 μm finish

PROBLEMS

8-1   Redraw the objects and dimension them in decimal inches. Scale the drawings for the dimensions.

a

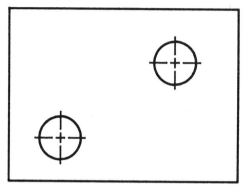

b

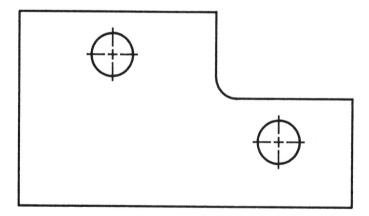

8-2 Redraw the objects and dimension them in millimeters. Scale the drawings for the dimensions.

a

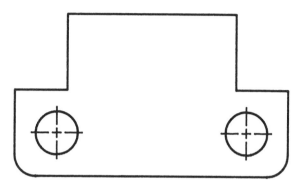

b

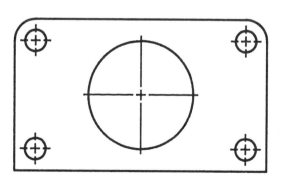

8-3   Redraw the objects and dimension them in millimeters. Scale the drawings for the dimensions.

a

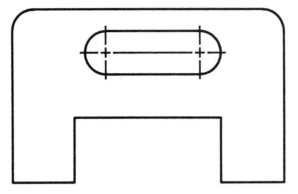

b

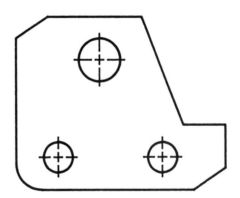

8-4 Redraw the object and dimension it in millimeters. Scale the drawing for the dimensions.

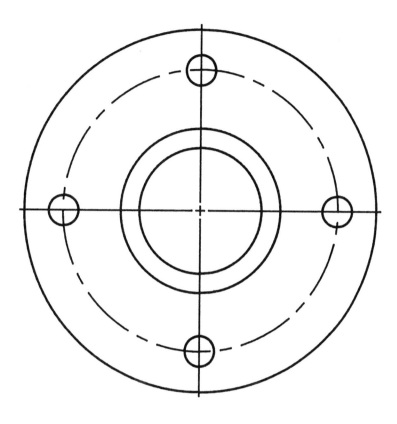

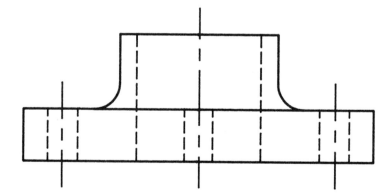

8-5 Redraw the object and dimension it in decimal inches. Scale the drawing for the dimensions.

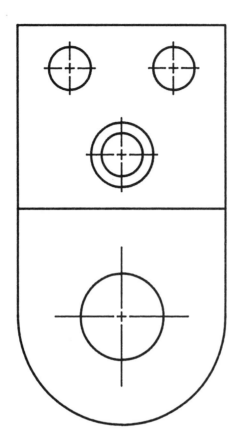
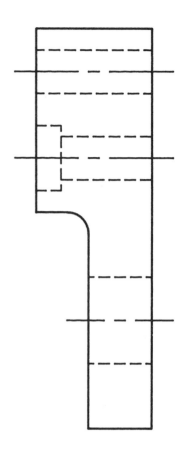

8-6 Redraw the object and dimension it in millimeters. Scale the drawing for the dimensions.

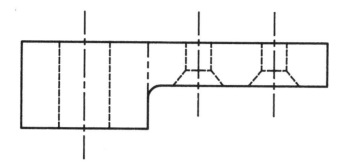

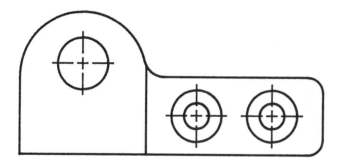

8-7 Redraw the object and dimension it in decimal inches. Scale the drawing for the dimensions.

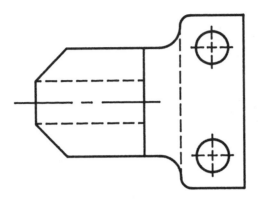

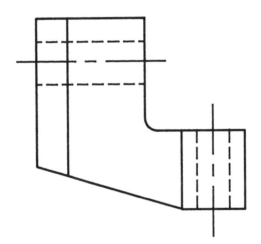

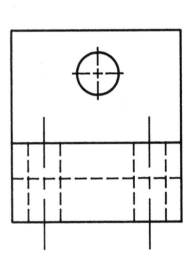

8-8 Refer to the drawing that follows and answer questions a through e.

   a. The value of the dimension marked X is $\underline{R\,1.19}$.

   b. The total length of the part is $\underline{2.89}$.

   c. The distance from the centerline of the pin holes to the pinion center, measured parallel to centerline sym, is $\underline{2.55}$.

   d. The tolerance on the pin holes is $\underline{.002}$.

   e. The distance from the gear shaft center to the end of the yoke is $\underline{3.14}$.

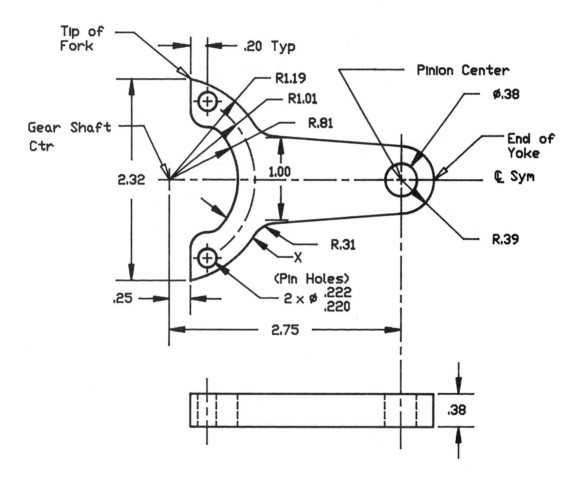

2.75
35
3.04

TIME ALARM.
APPLICATION FILED NOV. 16, 1907.

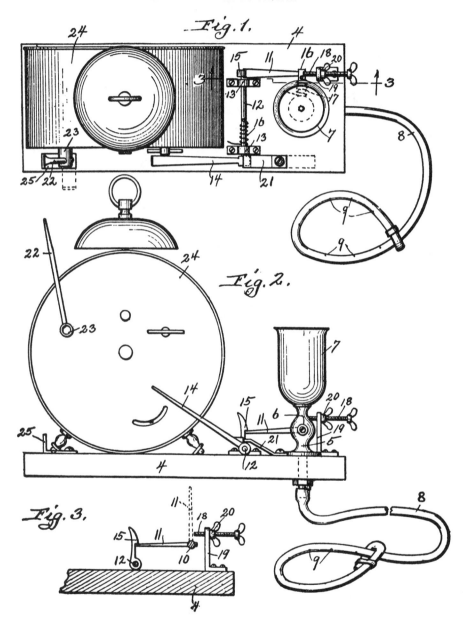

Fig.1.

Fig.2.

Fig.3.

WORKING DRAWINGS

9-1 INTRODUCTION

Have you ever wondered how ideas turn into products? For example, how does an idea for a device like a digital clock radio become a product ready for use? As you have probably guessed, it is a long and complex process. One of the first steps is design and the creation of several types of drawings.

The first types of drawings produced during the design process for any product or project are **design drawings,** sometimes called layouts. Making these drawings is a necessary step in the process, but the drawings do not contain all the information needed to build the end product. This information is contained in a type of drawing called a **working drawing.** The final step in the design process is the completion of the working drawing.

The illustration that opens this chapter shows a type of design drawing called a **patent drawing**. As you probably know, the federal government issues patents to protect inventions from theft. Inventors must apply for these patents, and part of the required application is a patent drawing. Patent drawings are different from any other type of mechanical drawings. For instance in the illustration that opens this chapter, notice the shading and lack of dimensions. Patent drawings are strictly design drawings because they do not contain enough information for building the invention. The particular invention pictured is an alarm device using a standard clock that, at a preset time, will open a valve and pour water onto a sleeping person. Needless to say, this invention never made it into production. The digital clock radio turned out to be a preferable alternative.

Working drawings are the principal means by which designers, engineers, and architects communicate their instructions to those who will build the end

After you complete this chapter, you should:

1. know the basic steps in and reasons for the drawing control process

2. know the six areas contained in a standard drawing format and the types of information contained in each

3. understand the use of zone coding on single-sheet and multi-sheet drawings

4. be able to recognize machining, casting, forging, and sheet metal detail drawings

5. recognize the differences between detail, assembly, and installation drawings

6. understand the uses of working assembly, subassembly, and overall assembly drawings

7. know the basic methods of fastening

8. recognize schematics and diagrams and understand their basic uses

9. recognize the fundamental differences between manufacturing and construction drawings

283

products. *Working* means that work can be done by using the information contained in these technical drawings. All working drawings do not look alike. There are a number of different types of working drawings and considerable variation in their appearance. For example, significant differences exist between construction plans, manufacturing drawings, and electronic schematics. The principal differences are in dimensioning methods, tolerance applications, symbol usage, and formats. This chapter covers the formats of detail, assembly and installation types of working drawings, and schematic drawings.

9-2 THE DRAWING CONTROL PROCESS

The creation of a working drawing can be a long and complicated process because of the need for continuous and accurate control of all the design documents. This control is exercised by the engineering or design department. The path from the original idea to the finished drawing is usually referred to as the drawing control process, and it generally includes the following steps:

1. Designers or engineers put their ideas down on paper in the form of a design drawing, or layout.

2. Drafters turn the layouts into working drawings, commonly done on vellum or Mylar.

3. Drawings are checked for mistakes and possible improvements, and changes are made as required.

4. When satisfactory, the drawings are signed by everybody involved in their production, including the drafter, the designer, and the engineering or design group supervisor. The drawing is then said to be released.

5. Copies of the drawing are made and distributed to all those who need them. The original of the drawing is stored in a safe place and always kept under the control of the engineering or design department. As an extra precaution, a microfilmed copy of the drawing is often made. All production work is done using copies called blueprints, or just prints.

6. For a variety of reasons, drawings frequently need to be changed. Customers change their minds, mistakes are discovered, and better designs are created. These changes—or revisions, as they are commonly called—are made to the original. New copies are printed and distributed, and the old ones are collected and destroyed.

9-3 WORKING DRAWING FORMATS

Working drawings contain a wide range of both detailed and general information. To make working drawings as easy as possible to use, this information is organized and shown according to a variety of standard conventions. These standard ways of showing information are called formats. The formats used on working drawings have the same purpose as the formats used in other documents, such as books and magazines. For example, books have tables of contents, chapters, page numbers, indices, and so on. In like manner, working drawings are divided into sections and contain tables and similar types of information. In other words, working drawings are not just pictures. It is important to note that the formats discussed in the following paragraphs include all information that *might* be found in a working drawing. The type of information displayed and the ways in which it is displayed may vary somewhat from one company to another. Also, formats vary depending on the drawing type (manufacturing or construction) and the drawing complexity (whether it is a detail, assembly, installation, or schematic drawing).

The information included in a drawing is usually found in one of six areas: the title block, parts list, notes, revision block, application block, or picture. The location of these areas varies somewhat, and some drawings may not contain all six. Figure 9-1 shows a typical format for a single-sheet drawing.

Figure 9-1 Typical single-sheet drawing format.

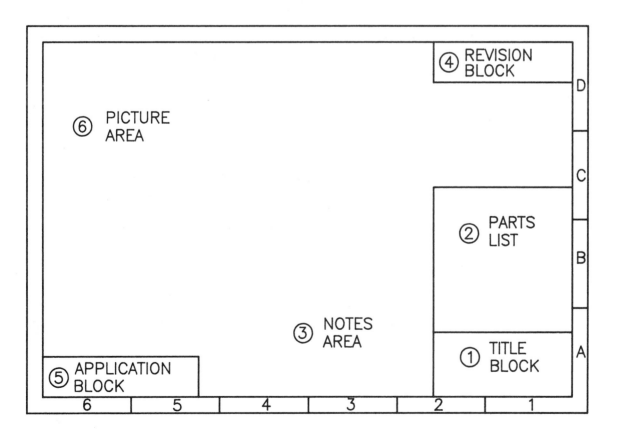

For multisheet drawings, it is common practice to reserve sheet 1 for the parts list, notes, and so on and to use subsequent sheets for pictures only.

Title Blocks

The title block is usually in the lower right-hand corner of the drawing and generally contains the following items:

Title: A descriptive noun phrase that usually contains the term *assembly* (or the abbreviation ASSY) or the term *installation* (or the abbreviation INSTL), if applicable.

Drawing number: A number that uniquely identifies each drawing. It may be a simple serial number or may signify certain information such as drawing sheet size, company division, type of product, or the like.

Sheet number: A drawing is not necessarily contained on a single sheet; many drawings contain multiple sheets, all of which have the same drawing number. Each sheet, however, has a different sheet number. The system most widely used indicates the number of the current sheet and the total number of sheets for the drawing. For example, the designation "Sh 1 of 5" means that the current sheet is sheet 1 of a total of five sheets.

Names: Consists of the name of the company that produced the drawing or the name of the originating organization.

Signatures: Includes all those responsible for creating or approving the drawing—the drafter, designer, engineer, and so on.

Scale: Indicates the scale of the picture.

Dimensional units and tolerance: Tells the units in which the drawing is dimensioned and the applicable tolerances.

Date: Drawing release date.

Materials: Specifications of the necessary materials, including their weight. These specifications may be included if no parts list is used.

Figure 9-2 shows several examples of typical title blocks.

Parts Lists

Parts lists are commonly located just above the title block. They are also called lists of materials, bills of materials, and schedules. Detail drawings

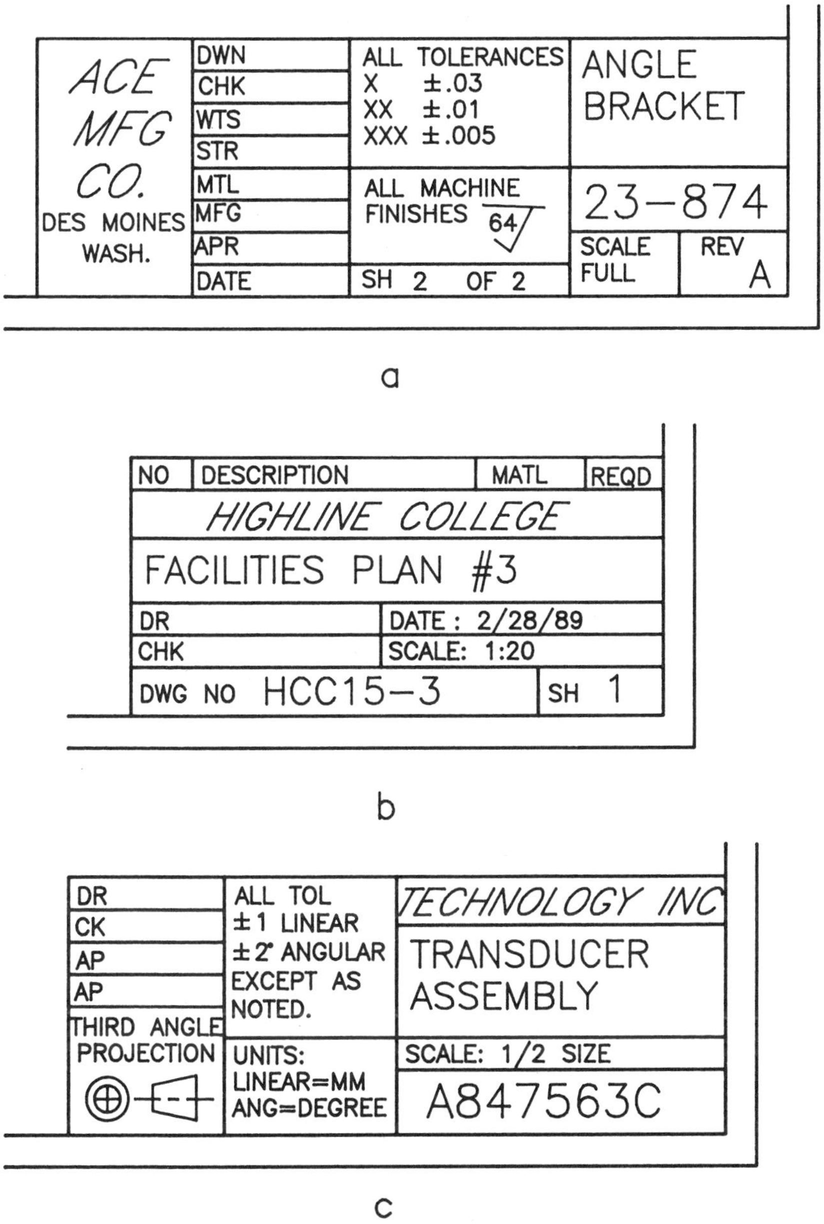

ACE MFG CO. DES MOINES WASH.	DWN	ALL TOLERANCES X ±.03 XX ±.01 XXX ±.005	ANGLE BRACKET	
	CHK			
	WTS			
	STR			
	MTL	ALL MACHINE FINISHES 64✓	23−874	
	MFG			
	APR		SCALE FULL	REV A
	DATE	SH 2 OF 2		

a

Figure 9-2 Examples of title blocks.

NO	DESCRIPTION	MATL	REQD
	HIGHLINE COLLEGE		
	FACILITIES PLAN #3		
DR		DATE : 2/28/89	
CHK		SCALE: 1:20	
DWG NO HCC15−3			SH 1

b

DR	ALL TOL ±1 LINEAR ±2° ANGULAR EXCEPT AS NOTED.	TECHNOLOGY INC
CK		
AP		TRANSDUCER ASSEMBLY
AP		
THIRD ANGLE PROJECTION ⊕-◁-⊣	UNITS: LINEAR=MM ANG=DEGREE	SCALE: 1/2 SIZE
		A847563C

c

sometimes do not contain a parts list. Parts lists vary considerably in appearance, but the following items are commonly included:

Part number

Part name

Quantity required

Material

Stock size

Part marking method

Finish requirements

Parts lists are typically read from bottom to top. That is, the first item needed is normally the one at the bottom and the last needed is at the top. Figure 9-3 illustrates several examples of typical parts lists.

Notes Areas

Notes are used on drawings to supply information that can best be presented in written form. There are two basic types: general and specific. General notes refer to some aspect of the whole drawing; specific notes refer to one area or feature. Specific notes are sometimes referred to as flag notes. They are used when a piece of information must be repeated several times or when there is no room to print the information needed (such as a lengthy material

2	GADGET		RUBBER	2
1	WIDGET		CI	5
NO.	PART NAME		MATERIAL	QTY
		PARTS LIST		

a

		X	1	4	KNEE	ICONEL		RS	A
			X	3	12	ELBOW	ICONEL		RS	
		−2−1	QTY	NO.	DESCRIPTION	MATERIAL	FINISH	PT MARK	REV
RELEASE					LIST OF MATERIAL				

b

25D538	WEBBED FLANGE	32x59x10	1▷	2▷	5
PART NO.	PART NAME	STOCK SIZE	MATERIAL	FINISH	QTY
		PARTS LIST			

c

Figure 9-3 Examples of parts lists.

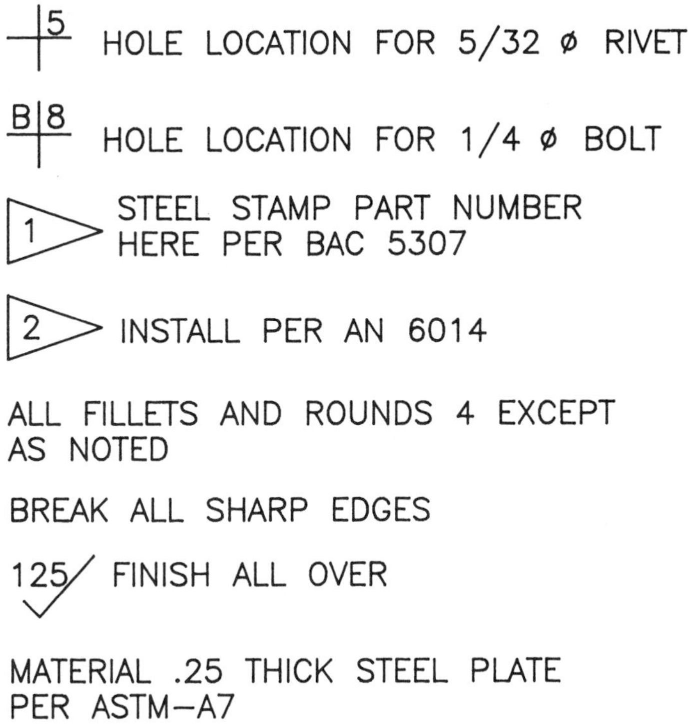

Figure 9-4 Examples of specific and general notes.

—|5 HOLE LOCATION FOR 5/32 ∅ RIVET

B|8 HOLE LOCATION FOR 1/4 ∅ BOLT

▷1 STEEL STAMP PART NUMBER
HERE PER BAC 5307

▷2 INSTALL PER AN 6014

ALL FILLETS AND ROUNDS 4 EXCEPT
AS NOTED

BREAK ALL SHARP EDGES

125/ FINISH ALL OVER
⌄

MATERIAL .25 THICK STEEL PLATE
PER ASTM—A7

callout in the parts list). For specific notes a flag such as ▷ or ▭▷ (other symbols are also used) is shown at the point of application and again in the notes area along with the printed information. Notes may be anywhere on a drawing. Figure 9-4 illustrates some typical notes.

Revision Blocks

Included in revision blocks are brief descriptions of any changes made to the drawing after its original release. Each change is assigned a letter (A, B, and so forth) to identify it. The letter designating the change is shown on the drawing in the places where the changes were made as well as in the change block. Also included in the block are dates and signatures. This block is typically located in the upper right-hand corner. Figure 9-5 shows several examples.

Application Blocks

With some exceptions, which will be explained later, there are two fundamental types of working drawings: detail (DET) and assembly (ASSY). Detail drawings show individual parts; assembly drawings show how detail parts are

SYM	REVISIONS	DATE	CK
A	.503 ⌀ HOLE ADDED	3/8/89	GCS
B	1.245 WAS 1.254	4/9/89	RM

a

REV	REVISION DESCRIPTION	APPD	DATE
A	ADDED FINISH NOTE	SRS	2/15/89
B	4 PLACES WAS 3 PLACES	TMS	5/4/89

b

Figure 9-5   Examples of revision blocks.

fastened together into groups of two or more. Simple assemblies require only a few detail drawings and one assembly drawing. Large and complex assemblies may require many detail and sub-assembly drawings, and an overall assembly drawing. Extremely large assemblies, such as 747 airplanes, have many thousands of detail and assembly drawings. All these drawings must be organized in such a way that users can find their way from detail drawings to subassembly drawings to more complex assembly drawings and back again. The organizing strategy is summarized by a *drawing tree.* Figure 9-6 shows an example.

The application block (sometimes called the used-on block) provides information about where a drawing fits in the drawing tree. It tells what the next "higher" drawing is and what model this detail or assembly is used on. Drawings that do not have an application block include this information in the title block or in a note. Figure 9-7 shows an example of an application block.

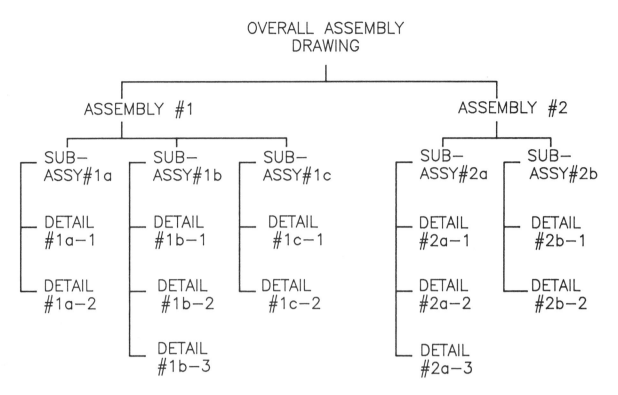

Figure 9-6 A typical drawing tree.

APPLICATION			
NEXT ASSY	MODEL USED ON	ASSY NO	REV LTR
24−7387	WD4−98	−1	
24−7388	SD3−04	−2	A

Figure 9-7 An example of an application block.

Picture Areas

The picture is, of course, an integral part of the drawing. On a single-sheet detail drawing, the picture is usually toward the left side of the sheet and is easily locatable. On a large and complex assembly drawing, the picture is likely to be spread over many sheets; therefore, it may be difficult to locate specific items. Figure 9-8 shows a drawing sheet with letters and numbers printed in the border. These are called zone codes. Each letter represents an imaginary row extending horizontally across the sheet. Each number repre-

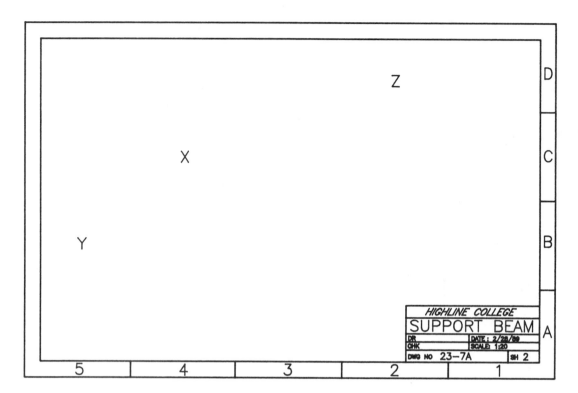

Figure 9-8 Drawing zones.

sents an imaginary column running vertically from the bottom to the top of the sheet. The intersection of a horizontal row and a vertical column is called a zone. Zones are used to locate objects and features on large drawings. In Figure 9-8 "X" has been placed opposite "C" and above "4." The location of "X" is referred to as zone C4. "Y" has been placed in zone B5 and "Z" in zone D2.

In Figure 9-9, section view indicators have been placed on the drawing sheet in two locations. Note that each has been labeled with the zone in which it appears. This is called labeling by zone. The advantage to this system is that users of the drawings always know where to look for the cutting plane lines when they want to know where the section view came from.

On very large drawings, cutting plane lines or view indicators are labeled as shown in Figure 9-10. Here, two zones are indicated: One is the location of the cutting plane indicator, and the other is the location of the section view. In the example, the cutting plane line is labeled "$C5_{B3}$." "C5" indicates the location of the cutting plane, and "$_{B3}$" shows the location of the section view. The section view is labeled "SECTION C5." Thus, the cutting plane refers to the location of the section view and the section view refers to the location

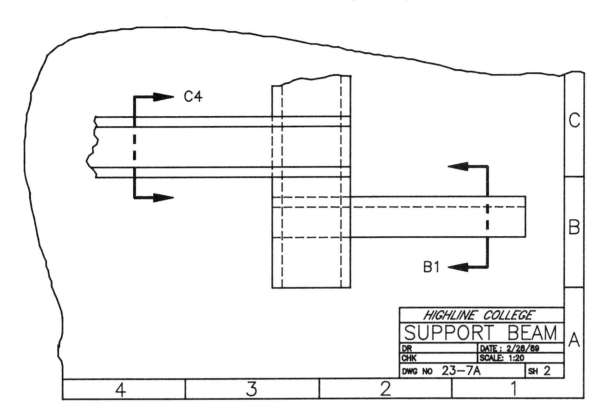

Figure 9-9 Labeling by zone.

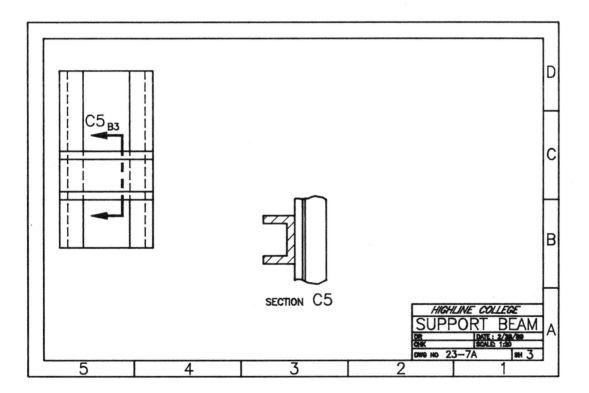

Figure 9-10 Labeling by zone on a large drawing.

of the cutting plane. In some cases the section view may be drawn on a different sheet than the cutting-plane view indicator. This information is included in a label such as "A5$_{SH3-B2}$."

Zone location information is frequently included in parts lists on complex multisheet drawings. This greatly simplifies the job of locating particular parts.

9-4 DETAIL DRAWINGS

A drawing used in the making or fabricating of a single part is called a **detail drawing,** or **DET.** Separate detail drawings are not necessarily made for each individual part because several parts may be detailed on one drawing. Detail drawings must include all the information needed to make individual parts. Detail information includes:

Size (dimensional) and shape (pictorial) description

Material and heat-treating requirements

Protective finish instructions (painting, plating, and the like)

Surface texture requirements

Part numbering and marking instructions

The next higher drawing number (what assembly the part will be used in)

Quantity required

Machining Drawings

Drawings showing detail parts cut or machined out of solid material are usually the simplest type of detail drawing. Figure 9-11 shows an example of a fairly simple detail drawing that has only a picture and a title block containing the material and dimensional tolerance callouts.

Figure 9-12 shows a more complex detail drawing. The object itself is relatively simple, but the manufacturing instructions and the drawing format are more complicated. Included on the drawing are blocks for applications (lower left-hand corner), revisions (upper right-hand corner), and a parts list, in addition to the title block. Notice that a list of production standards notes is also included.

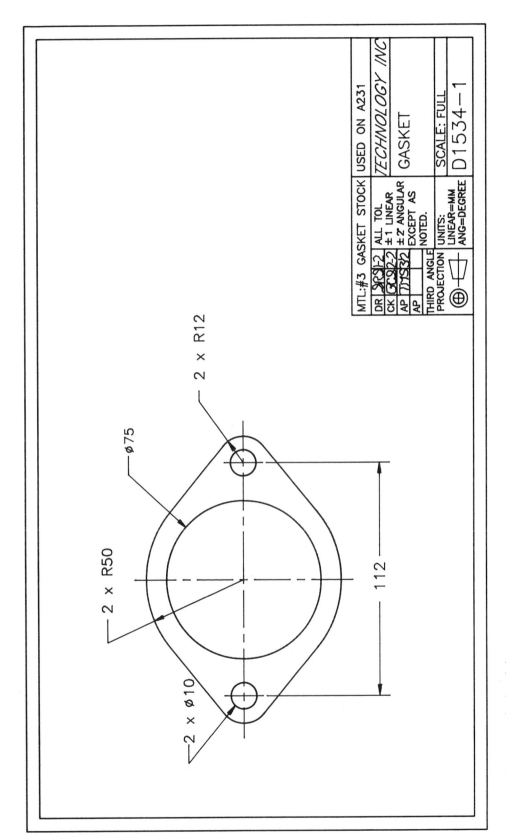

Figure 9-11 A simple detail drawing.

295

Figure 9-12 A relatively complex detail drawing.

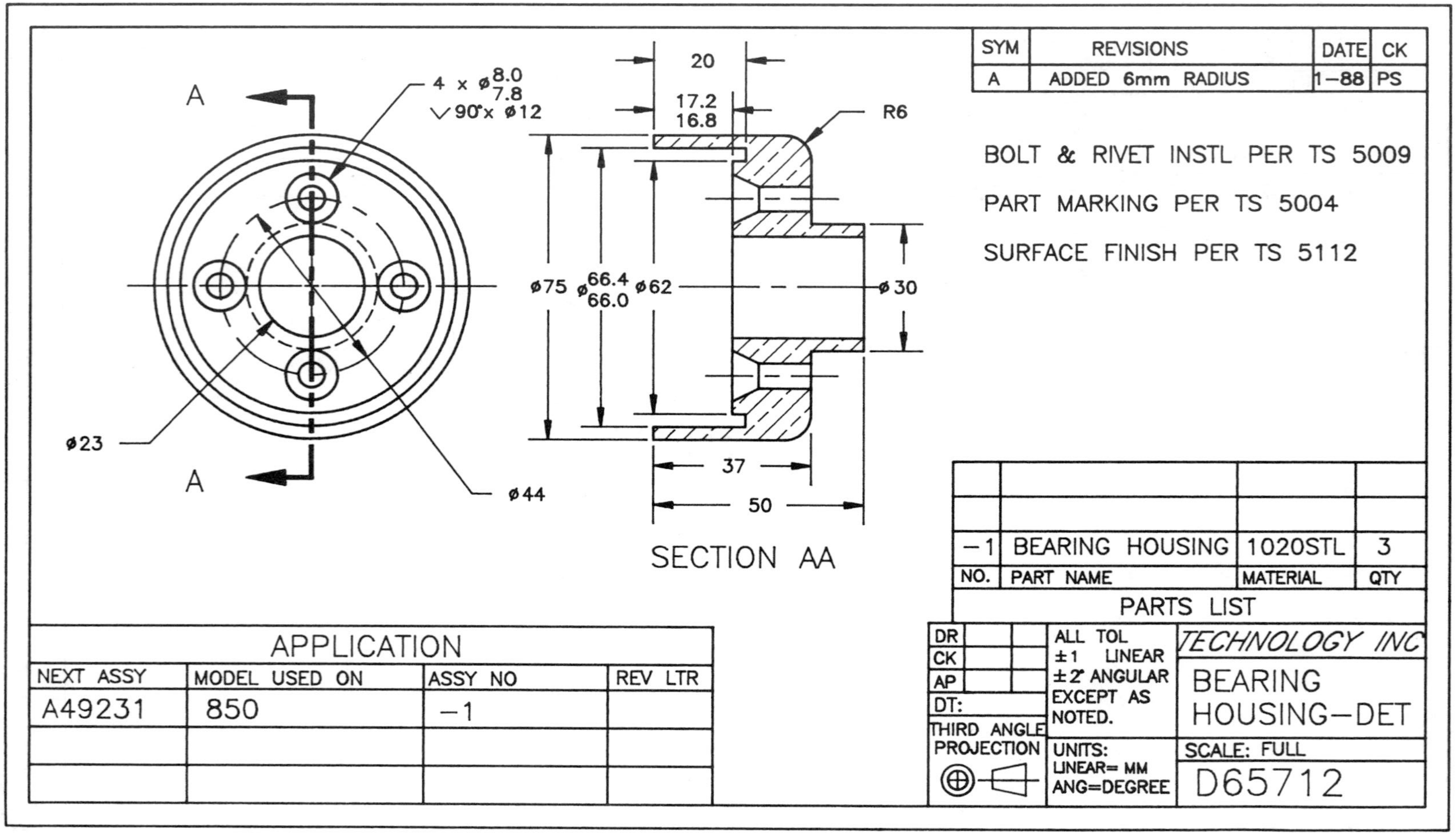

SYM	REVISIONS	DATE	CK
A	ADDED 6mm RADIUS	1-88	PS

BOLT & RIVET INSTL PER TS 5009

PART MARKING PER TS 5004

SURFACE FINISH PER TS 5112

NO.	PART NAME	MATERIAL	QTY
-1	BEARING HOUSING	1020STL	3

PARTS LIST

TECHNOLOGY INC

BEARING HOUSING-DET

SCALE: FULL

D65712

ALL TOL ±1 LINEAR ±2 ANGULAR EXCEPT AS NOTED.

UNITS: LINEAR= MM ANG=DEGREE

DR
CK
AP
DT:
THIRD ANGLE PROJECTION

R6
Ø30
20
17.2
16.8
50
37
Ø62
Ø66.4
Ø66.0
Ø75
Ø44
Ø23
4 x Ø8.0 / Ø7.8
⌵90°x Ø12

SECTION AA

APPLICATION

NEXT ASSY	MODEL USED ON	ASSY NO	REV LTR
A49231	850	-1	

Casting and Forging Details

Casting and forging are both methods of forming parts through working hot metals. In casting, metal is heated until molten and then poured or forced into a mold. When the metal cools, it takes the shape of the mold cavity. There are many different casting processes, but drawings related to casting are all basically the same. In forging, metal is heated to a plastic state and then forced into the required shape by placing it between two dies and applying tremendous force.

Except for a few of the casting processes, such as die casting, the surface finishes and dimensional control of parts produced by casting and forging are not of very high quality. To achieve smooth surfaces and close tolerance control, additional machining operations are required.

If the parts are simple, casting or forging information may be combined on a single drawing with the machining information. The cast or forged material that is to be removed by machining is shown by phantom lines. If complex, casting and forging information may appear on separate drawings. If a separate machining drawing is made, dimensions and other information that does not change are not repeated. The rough and finished parts may have separate part numbers. In this case, the machining drawing must indicate that the finished part is made from the forging or casting. See Figure 9-13 for an example of a combined casting and machining drawing.

Figure 9-13 A combined casting and machining detail drawing.

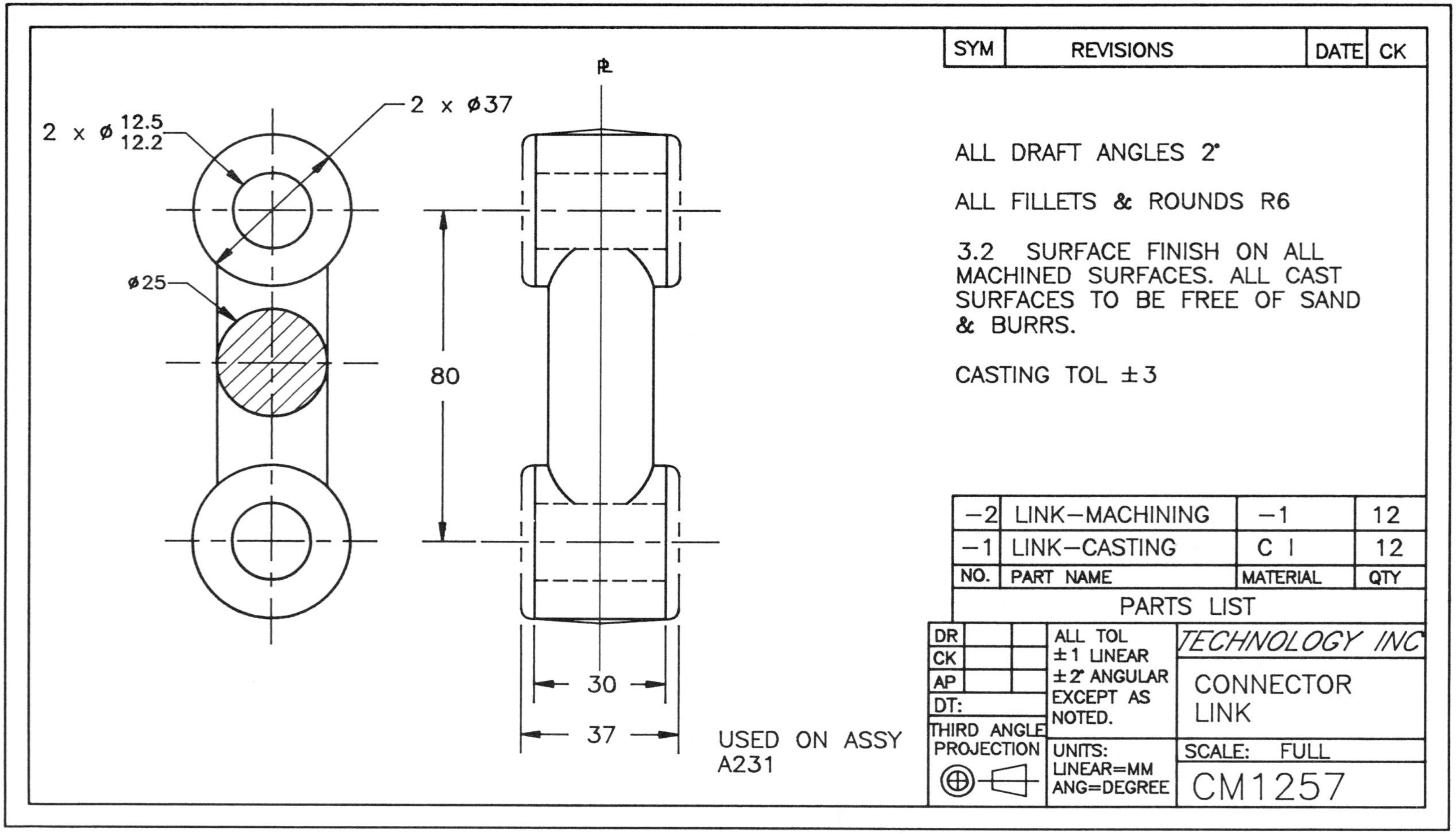

ALL DRAFT ANGLES 2°

ALL FILLETS & ROUNDS R6

3.2 SURFACE FINISH ON ALL MACHINED SURFACES. ALL CAST SURFACES TO BE FREE OF SAND & BURRS.

CASTING TOL ±3

SYM	REVISIONS	DATE	CK

2 x Ø37

2 x Ø 12.5 / 12.2

Ø25

80

30

37

USED ON ASSY A231

NO.	PART NAME	MATERIAL	QTY
-2	LINK-MACHINING	-1	12
-1	LINK-CASTING	C I	12

PARTS LIST

TECHNOLOGY INC

DR		ALL TOL ±1 LINEAR ±2 ANGULAR EXCEPT AS NOTED.	CONNECTOR LINK	
CK				
AP				
DT:				
THIRD ANGLE PROJECTION		UNITS: LINEAR=MM ANG=DEGREE	SCALE: FULL	CM1257

Sheet Metal Details

A part made from sheet metal may be drawn as a standard multiview showing the finished product or as a development showing the flat pattern. Figure 9-14 shows a drawing of a formed sheet-metal part. This part started as a flat sheet of metal out of which the shape was cut. It was then bent, or formed, into its final shape as shown by the drawing. Because no angular dimensions are given, it is assumed that all bends are 90°.

Figure 9-15 shows examples of sheet metal details drawn as flat patterns. In these examples the shape of the flat sheet is shown before it is bent into its final form. Notes or callouts such as "BEND UP 90° × .19R" indicate the direction of the bend (up), amount of bend (90°), and the radius (.19) for the bend.

Notice that three different parts are detailed in Figure 9-15. Only two different parts are shown in the picture, but three are indicated in the parts list. They are identified as –1, –2, and –3. The full part numbers include the drawing number, 489FP2; the dash; and the number that follows the dash. In other words, the part numbers are 489FP2−1, 489FP2−2, and 489FP2−3. The drawing shows pictures of –1 (the hanger bracket) and –3 (the angle bracket), but –2 is listed simply as "OPP−1," which means the opposite of −1.

300

Figure 9-14 A drawing of a formed sheet-metal part.

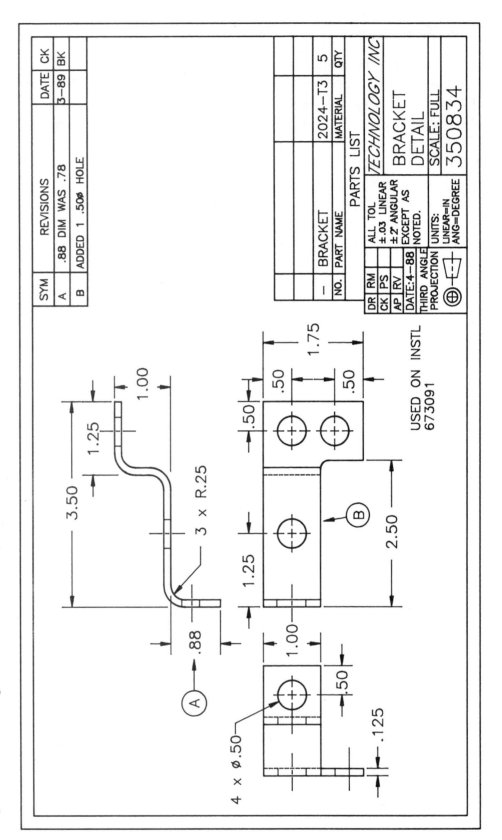

SYM	REVISIONS	DATE	CK
A	.88 DIM WAS .78	3–89	BK
B	ADDED 1 .50φ HOLE		

PARTS LIST

NO.	PART NAME	MATERIAL	QTY
–	BRACKET	2024–T3	5

TECHNOLOGY INC

DR	RM	ALL TOL	BRACKET
CK	PS	±.03 LINEAR	DETAIL
AP	RV	±2° ANGULAR EXCEPT AS NOTED.	

DATE:4–88

THIRD ANGLE PROJECTION

UNITS: LINEAR=IN ANG=DEGREE

SCALE: FULL

350834

USED ON INSTL 673091

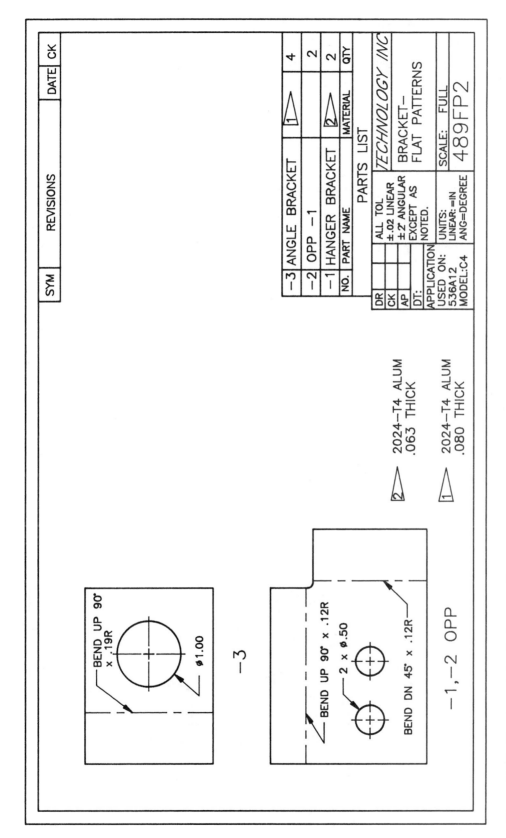

Figure 9-15 Sheet metal details drawn as flat patterns.

301

Opposite hand parts are very common on symmetrical assemblies such as aircraft and automobiles. They are not identical parts, but mirror images, like a pair of shoes. The left wing of an airplane is a mirror image of the right wing. On production drawings it is common practice to draw only one side and indicate in a note that the other side is an opposite. Figure 9-16 shows another example of opposite parts.

9-5 ASSEMBLY DRAWINGS

An **assembly** is sometimes defined as a collection of parts that performs some independent function. Some everyday examples are pencil sharpeners, calcu-

Figure 9-16 Opposite parts are mirror images of each other.

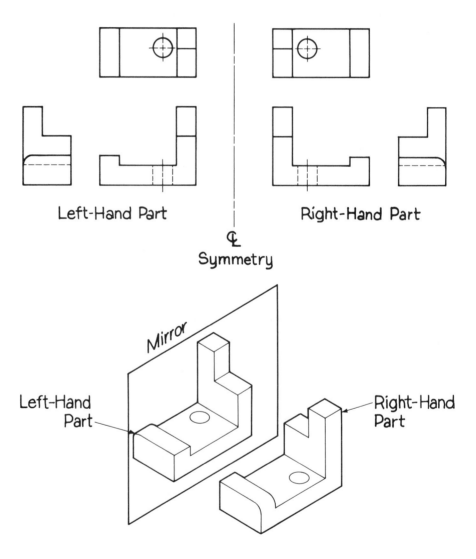

lators, and bicycles. But assemblies such as automobile engines and home-heating furnaces are part of larger assemblies. They perform functions, but not independently of the functions of cars or houses.

An **assembly drawing,** or **ASSY,** shows groups of parts that mate, or fit, with each other in a permanent, or semipermanent configuration. The purposes of assembly drawings are to show the relationships between parts and the methods by which they are fastened together—in other words, to show how assemblies are to be built. Assembly drawings usually provide the following information:

A picture showing how the detail parts fit together

A list of required component parts

Location dimensions

Fastening methods

A part number for the finished assembly

The next higher drawing number (the number of the drawing that shows the structure in which the assembly will be used)

An assembly drawing may provide detailed fabrication information about some of or all the parts from which it is made. If an assembly and its parts are simple enough, a drawing may be made that shows both the detail and the assembly information on the same drawing. This combination detail-assembly drawing is called a working assembly. On the other hand, when it becomes impractical to show a large, complex assembly on one drawing, the drawing of the assembly may be broken up into several smaller groupings called **subassemblies.** Separate drawings are made of each subassembly, and these are then combined in an overall assembly.

It is important to note that overall assembly drawings do not repeat all the information shown on detail or subassembly drawings. An attempt is made to simplify and thus clarify assembly drawings by eliminating as many hidden lines and confusing detail as possible. Assembly pictures include only enough detail to show the individual parts or subassemblies. There is a great difference between detailing parts and merely showing parts. Detailing parts requires complete picture descriptions, but to show parts requires that only the outlines of the parts be pictured. The only dimensions shown on standard assemblies are those relating to locations of parts or movement. Parts can also be shown by means of symbols, incomplete pictures, or even by locations only. Exceptions to these practices are found, of course, on working assembly drawings.

Despite the simplicity of part description, assembly drawings can be very complex and difficult to read. For this reason extensive use is made of section views and partial views. The detail parts of an assembly are identified on the picture by use of item numbers, or part numbers. Figure 9-17 shows a drawing in which each part is identified by an item number. These item numbers are listed in the parts list, which gives the part number, description, and quantity for each part. Notice that the drawing is quite simple. It contains only two views, no dimensions, and very few notes.

Figure 9-18 shows an example in which detail drawing numbers, instead of item numbers, are used on the picture and in the parts list. These part numbers refer to particular detail drawings and standards documents. If you needed to find the dimensions of part HC-257, you would have to look at detail drawing HC-257.

Figure 9-19 shows an example of a working assembly drawing. All the information required for making the detail parts, including dimensions and material specifications, plus all the information needed to assemble the detail parts are included on this one drawing.

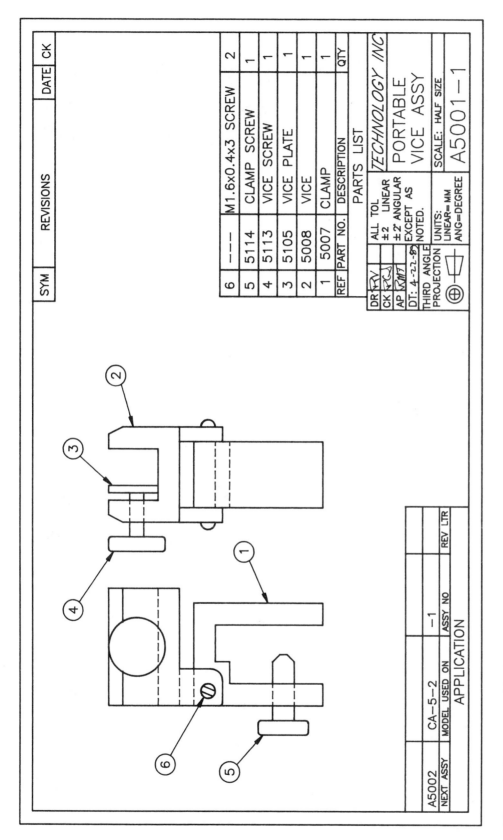

SYM	REVISIONS	DATE	CK

6	---	M1.6x0.4x3 SCREW	2
5	5114	CLAMP SCREW	1
4	5113	VICE SCREW	1
3	5105	VICE PLATE	1
2	5008	VICE	1
1	5007	CLAMP	1
REF	PART NO.	DESCRIPTION	QTY

PARTS LIST

DR	NV	
CK	RC	
AP	RMT	
DT: 4-22-8		

THIRD ANGLE PROJECTION

ALL TOL
±2 LINEAR
±2° ANGULAR
EXCEPT AS
NOTED.

UNITS:
LINEAR=MM
ANG=DEGREE

TECHNOLOGY INC

PORTABLE
VICE ASSY

SCALE: HALF SIZE

A5001-1

| A5002 | CA-5-2 | -1 | |
| NEXT ASSY | MODEL USED ON | ASSY NO | REV LTR |

APPLICATION

Figure 9-17 An assembly drawing with item numbers used to identify individual detail parts.

305

Figure 9-18 An assembly drawing in which detail drawing numbers identify parts.

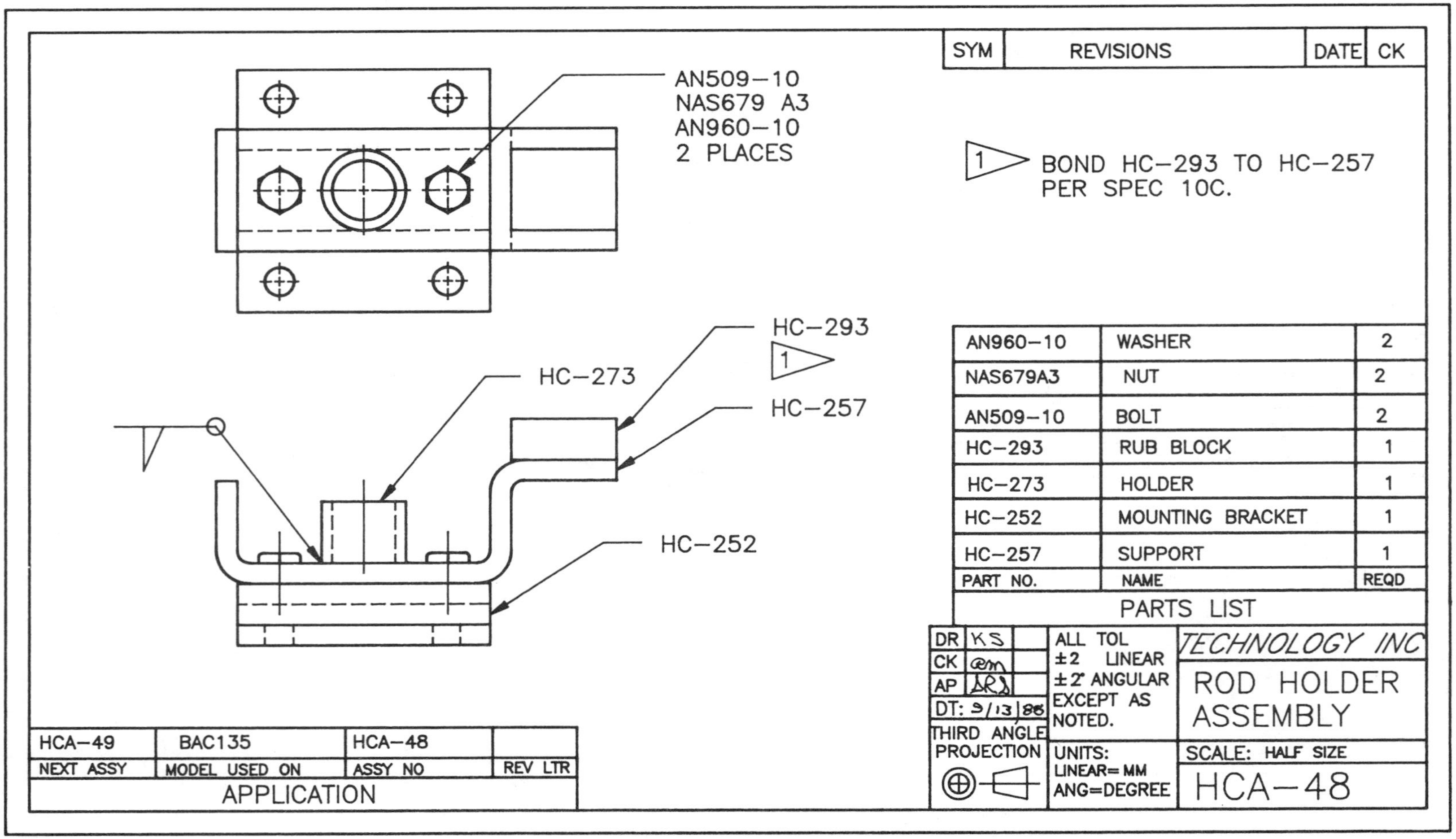

PART NO.	NAME	REQD
AN960-10	WASHER	2
NAS679A3	NUT	2
AN509-10	BOLT	2
HC-293	RUB BLOCK	1
HC-273	HOLDER	1
HC-252	MOUNTING BRACKET	1
HC-257	SUPPORT	1

PARTS LIST

SYM	REVISIONS	DATE	CK
1	BOND HC-293 TO HC-257 PER SPEC 10C.		

TECHNOLOGY INC
ROD HOLDER ASSEMBLY
SCALE: HALF SIZE
HCA-48

DR	KS
CK	@m
AP	AR

DT: 5/13/86

THIRD ANGLE PROJECTION

ALL TOL
±2 LINEAR
±2 ANGULAR
EXCEPT AS NOTED.

UNITS:
LINEAR=MM
ANG=DEGREE

AN509-10
NAS679 A3
AN960-10
2 PLACES

HC-293
HC-257
HC-273
HC-252

HCA-49	BAC135	HCA-48	
NEXT ASSY	MODEL USED ON	ASSY NO	REV LTR

APPLICATION

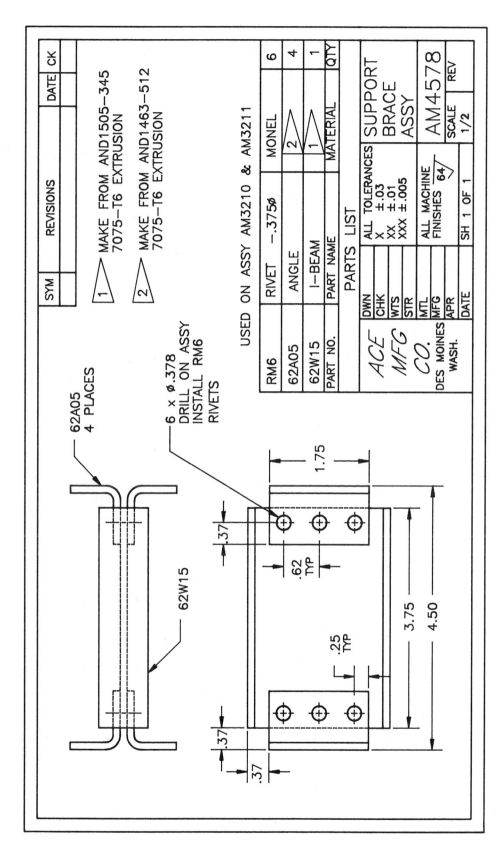

Figure 9-19 A working assembly drawing.

307

9-6 INSTALLATION DRAWINGS

An **installation drawing,** or **INSTL,** describes where and how details or assemblies are to be permanently installed. Installation drawings provide location dimensions (in reference to the entire structure) and fastening methods. Installation drawings are very similar to assembly drawings. The major difference between the two lies in the interpretation of the word *location.* Assembly drawings give locations of parts only as they pertain to each other; installation drawings give locations within the entire structure. Installation drawings are used for very large structures such as aircraft, ships, and buildings. Installation drawings describe where certain parts or assemblies are to be permanently affixed. Installation drawings provide the following information:

A picture showing details or assemblies in installed position

A list of required component parts

Location dimensions (in reference to the entire structure)

Fastening methods

A part number for the finished installation

The next higher drawing number (the number of the drawing that shows the structure that contains the assembly)

Note that the preceding items are almost identical to those listed for assembly drawings. Figure 9-20 shows an example of a typical installation drawing. Notice that the existing structure is shown by phantom lines; only the installed parts are shown with solid lines. Installation drawings are so similar to assembly drawings that the two are often combined. Detail information about the assembly or about installation components may also be given on the installation drawing. Thus, some installation drawings are actually detail-assembly-installation drawings.

9-7 FASTENING METHODS

Assemblies and installations require some method of fastening parts together. A wide range of methods are available, but they can be divided into two major categories: removable and permanent.

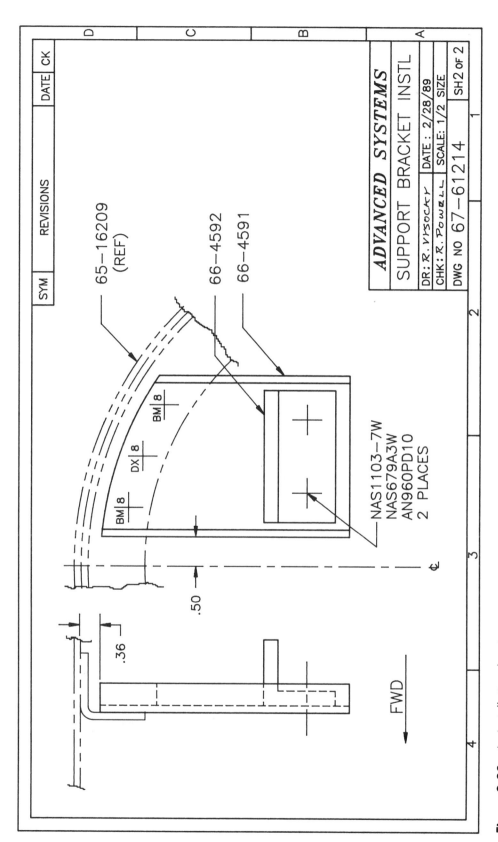

Figure 9-20 An installation drawing.

Removable Fasteners

Fasteners that can be installed and removed without destroying them include bolts, screws, and nuts. In most cases these will be standard parts used for economy as well as interchangeability.

Standard fasteners are identified on drawings by unique part numbers that are coded to indicate size, thread type, and material. Fasteners are seldom detailed on production drawings. Rather, symbols or centerlines are used to show the location; coded part numbers indicate the type and size. For example, in Figure 9-20 the two NAS1103-7W (bolts), NAS679A3W (nuts), and AN960PD10 (washers) are indicated by crossed lines. The detail information for them is contained in a standard parts catalog.

Permanent Fasteners

Permanent fasteners are those that must be destroyed to remove them. This category includes rivets and various types of locking fasteners. In addition, welding, soldering, and bonding are considered permanent methods of fastening. Rivets, like removable fasteners, are usually indicated by symbols and identified by callouts or codes as shown in Figure 9-20, in which "BM8" and "DX8" refer to types of rivets.

Welds are also indicated by a system of symbols, but the system is much more complex than that used for fasteners. Figure 9-21 shows some typical welded joints, and Figure 9-22 shows some applications of weld symbols on an assembly drawing. Table A-8 in the appendix presents weld symbols.

9-8 SCHEMATICS AND DIAGRAMS

Schematics and diagrams are simplified drawings of complex systems. They use symbols, lines, and reference designations to represent real objects in some cases and abstract concepts in others. In other words, a schematic does not look like the real thing, if there is a real thing. Obviously, schematics are dimensionless and cannot be drawn to any scale. However, the symbols are usually drawn to a standard size, and distances between lines are constant.

Schematics and diagrams are typically used in electrical, electronic, fluid-power, and plumbing design work. Figure 9-23 shows an example of an electronic schematic. The various symbols are used to represent electronic parts or components such as resistors and transistors. The reference designations ("R1," "Q1," and the like) act as part numbers. The lines connecting the symbols are not wires, but signal paths. The real physical assembly of these parts probably does not resemble the schematic at all.

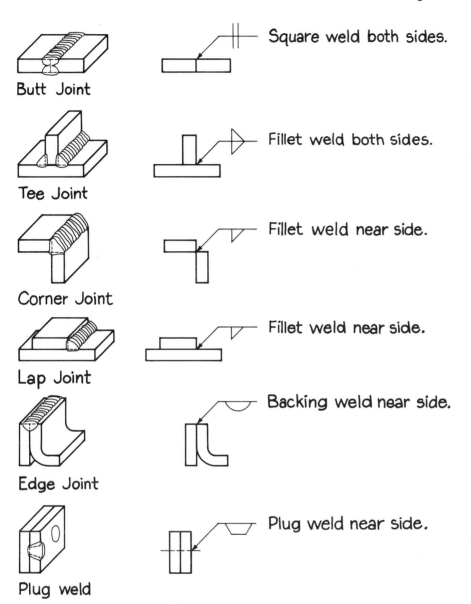

Figure 9-21 Typical welded joints and their associated weld symbols.

Figure 9-22 An assembly drawing that includes weld symbols.

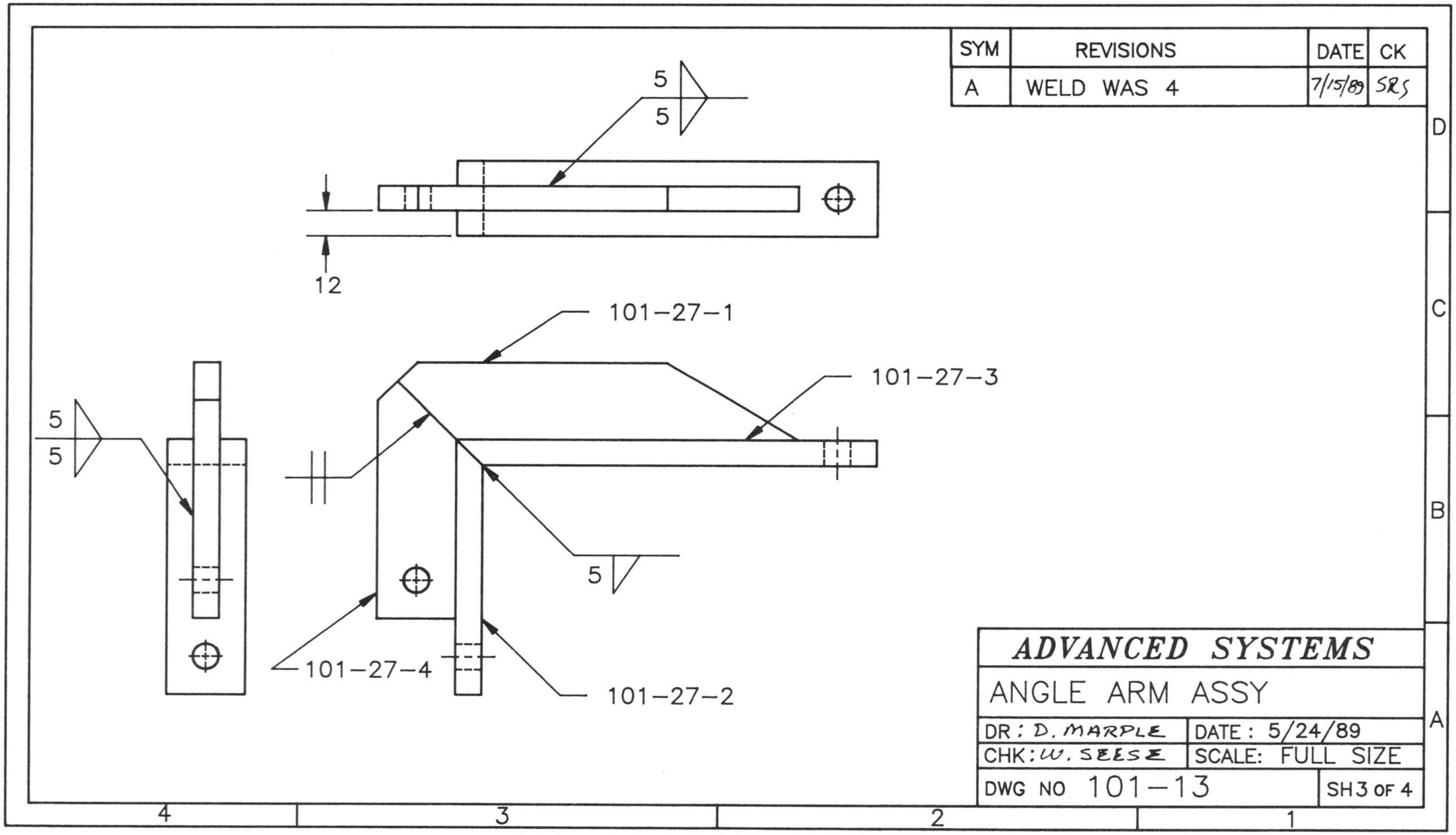

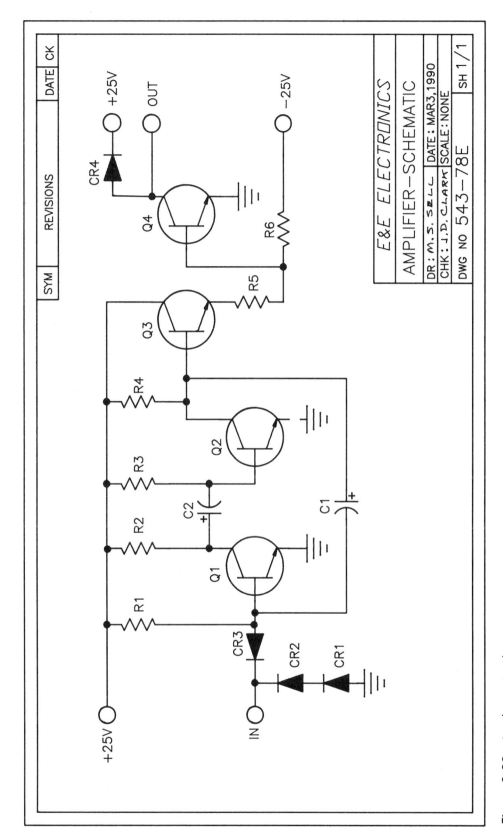

Figure 9-23 An electronic schematic.

313

An electrical wiring diagram is shown in Figure 9-24. Wiring diagrams are more realistic than schematics. The components are shown not by symbol, but by a very simplified representation of the part. The lines, in this case, do represent wires. Their length and routing are only approximate, but their destinations are indicated.

Compare the fluid-power diagram in Figure 9-25 with the plumbing diagram in Figure 9-26. The former is a schematic using symbols for components and lines for fluid pathways. The plumbing diagram also uses symbols for components, but the lines represent actual pipes and could, in fact, be dimensioned. In other words, the plumbing system could be built by using this diagram, but it would be difficult to construct the fluid-power system by using only its diagram.

The appendix contains several tables of schematic symbols.

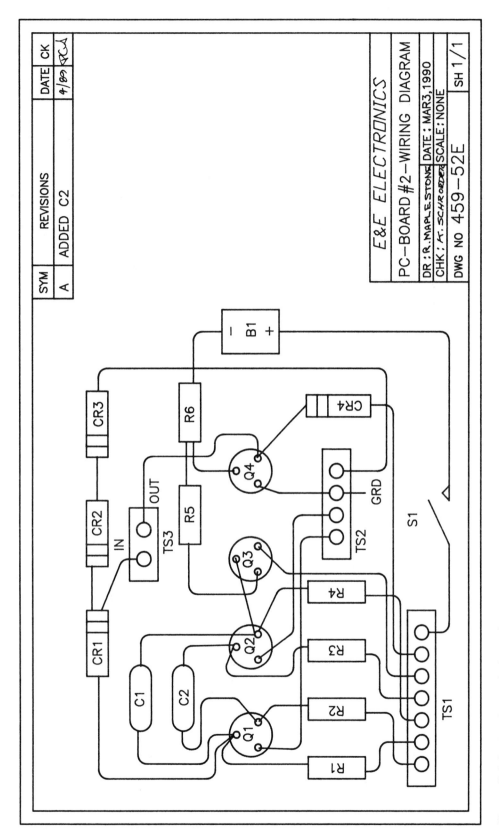

Figure 9-24 An electrical wiring diagram.

315

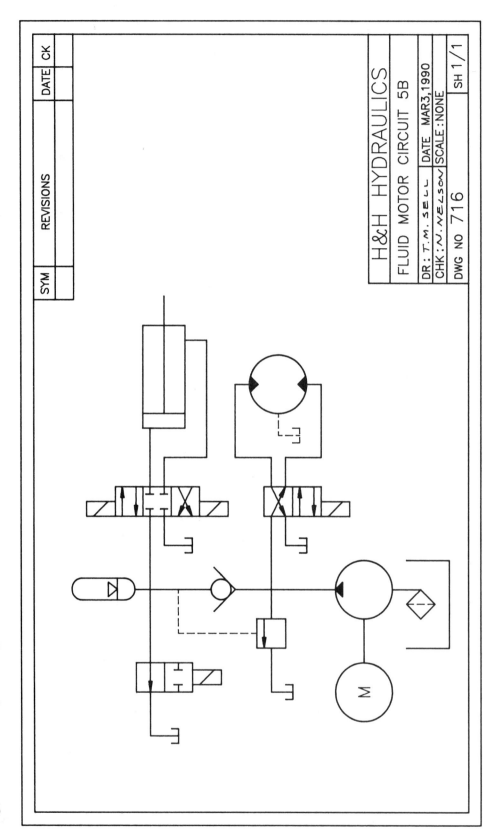

Figure 9-25 A fluid-power diagram.

316

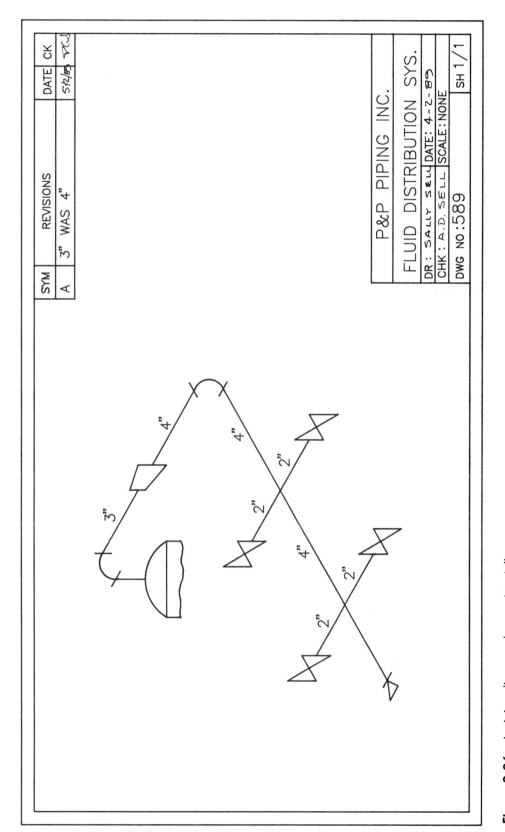

Figure 9-26 A piping diagram drawn pictorially.

317

9-9 CONSTRUCTION DRAWINGS

A construction drawing, or architectural drawing, includes all the previously discussed types of drawings. They usually contain detail, assembly, and installation information shown both schematically and pictorially. Construction drawings (often called plans) contain information about the structures, systems (electrical, plumbing, and so on), and envelopes (siding, roofing, windows, and the like) of buildings. A complete set of plans contains all the information necessary for the complete construction of a particular building. A set of plans for a small simple building may contain only a few sheets, but a set of plans for a large complex building contains hundreds of sheets.

Figure 9-27 shows a construction drawing for a garage. This is a relatively simple structure, and so the drawing is also simple. However, it does contain the basic format used on all construction drawings. There is a plan view, an elevation view (section A−A), and several detail views drawn to a scale larger than the main views. The structure is shown in all views. The envelope is described in the plan view and section A−A. The wiring (two lights and a switch) is shown schematically in the plan view. All the basic information needed for building the garage is contained in this one drawing.

9-10 PRINTS AND REPRODUCTIONS

Technical drawings are reproduced in several ways. They are copied on microfilm; as blueprints; and as whiteprints, or bluelines. The diazo process is the most common method for creating whiteprints and is the most widely used form of drawing reproduction. The diazo print is a reproduction of an original drawing; the print has dark lines and a light background. The diazo form of reproduction is a dry chemical process that utilizes a special paper coated with photosensitive chemicals, an ammonia developer, and an ultraviolet light source.

Diazo prints are made by placing the original drawing, created on a translucent media (such as vellum or Mylar), face up on top of the diazo paper and then exposing it to ultraviolet light. As the light passes through the translucent original and onto the diazo paper, it exposes the chemical coating. The chemical remains on the diazo paper only where the line work and lettering on the original drawing protect it from the light. The diazo paper is then separated from the original drawing and developed by passing it through ammonia vapor. The ammonia vapor causes the unexposed areas to turn dark, thus creating a positive reproduction with dark lines on a light background.

Though still the most common method for making whiteprints, the diazo process is slowly being replaced by xerographic copying.

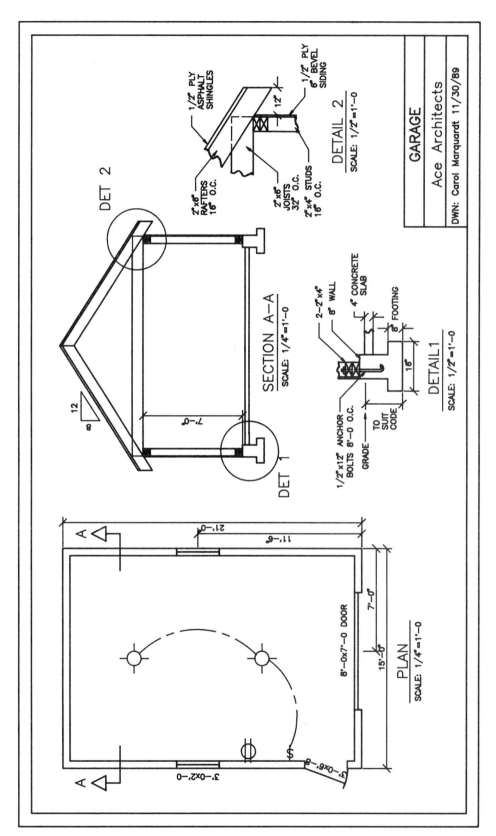

Figure 9-27 A construction drawing of a garage.

319

The following are guidelines for the effective use of prints:

- Don't ever measure a print to get a dimension. Distances on prints are never accurate because print paper is not dimensionally stable; it easily stretches and shrinks. Also, drawings are often drawn or printed in a reduced size. Rely on the dimensions given. Any dimension you might get from measuring the print is sure to be inaccurate.

- Take good care of prints. Don't eat lunch on them, walk on them, or use them with dirty hands. In other words, keep prints as clean as possible by using common sense in handling them.

- Be sure you are using up-to-date prints. Check periodically to see if there are later revisions.

9-11 CONCLUSION

This chapter has provided an introduction to working drawings. You have seen examples of detail, assembly, working assembly, and installation drawings as well as schematics, diagrams, and construction drawings. Every manufacturer or organization that produces working drawings uses its own unique format and sets its own drawing requirements. The examples included in this chapter are only representative of what exists in industry.

Check your general understanding of this chapter by answering the following review questions.

REVIEW QUESTIONS

1. *Layout* is another term for:
   a. a working drawing
   b. an assembly drawing
   c. a design drawing
   d. a detail drawing

2. After the drawing has been released, the shop works from the:
   a. original drawing
   b. copies of the drawing
   c. blueprints
   d. both b and c

3. Sheet numbers are usually found in the:

   a. title block

   b. parts list

   c. notes

   d. revision block

   e. application block

4. A brief description of a change made to the drawing after its initial release would be found in the:

   a. title block

   b. parts list

   c. notes

   d. revision block

   e. application block

5. The quantity of any parts is usually found in the:

   a. title block

   b. parts list

   c. notes

   d. revision block

   e. application block

6. Information telling the assembly in which a subassembly is used is usually found in the:

   a. title block

   b. parts list

   c. notes

   d. revision block

   e. application block

7. Consider a section view drawn in zone D5 and taken from a part shown in zone B2. Its cutting plane is labeled:

   a. B2

   b. D5

   c. B2-D5

   d. D5-B2

8. Which of the following types of information is a detail drawing *not* likely to contain?

   a. material

   b. part-numbering methods

   c. fastening methods

   d. size and shape

9. Which of the following is not a type of detail drawing?

    a. machining

    b. casting

    c. forging

    d. schematic

10. Which of the following types of assembly drawings contain fabrication (detail) information?

    a. subassembly

    b. working assembly

    c. overall assembly

    d. installation assembly

11. Installation drawings are used:

    a. in very large structures

    b. in very small structures

    c. in the exact same way as assembly drawings

    d. both b and c

12. Which of the following is considered a permanent type of fastener?

    a. screw

    b. rivet

    c. bolt

    d. snap ring

PROBLEMS

9-1   Refer to Figure 9-11 and answer the following questions:

    a. What is the drawing number?

    b. What is the name of the part?

    c. What is the drawing scale?

    d. What material is the part made from?

    e. What assembly is this part used in?

    f. What units is the part dimensioned in?

    g. What is the linear tolerance?

    h. What is the diameter of the small holes?

    i. What is the overall length of the part?

    j. How many holes are there in the part?

9-2   Refer to Figure 9-12 and answer the following questions:

    a. What is the drawing number?

    b. What is the name of the part?

    c. What is the drawing scale?

    d. What material is the part made from?

    e. What assembly is this part used in?

    f. What units is the part dimensioned in?

    g. What is the linear tolerance?

    h. How many small holes are there? What is the tolerance for them? What kind of holes are they?

    i. What is the diameter of the center hole?

    j. What is the overall length of the part?

9-3   Refer to Figure 9-13 and answer the following questions:

    a. What is the drawing number?

    b. What is the name of the company?

    c. What is the drawing scale?

    d. What material is the –1 part made from?

    e. What assembly is this part used in?

    f. What units is the part dimensioned in?

    g. What is the angular tolerance?

    h. What surface finish is specified for all machine surfaces?

    i. What is the diameter of the middle connecting cylinder?

    j. What is the overall length of the part?

    k. When was this drawing released?

    l. How much material is to be removed by machining from each side of the end cylinders?

9-4   Refer to Figure 9-14 and answer the following questions:

    a. What is the drawing number?

    b. What is the name of the part?

    c. What type of projection is used in this drawing?

    d. What material is the part made from?

    e. How many revisions has this drawing had?

    f. What units is the part dimensioned in?

    g. What is the linear tolerance?

    h. What change was made to create revision B?

    i. How many holes are there?

    j. What is the material thickness?

    k. What is the distance from the edge of either of the two right-hand holes to the edge of the part?

9-5 Refer to Figure 9-15 and answer the following questions:

    a. What is the drawing number?

    b. What is the name of the part?

    c. What model are these parts used in?

    d. What material are the parts made from?

    e. What assembly is this part used in?

    f. What is the thickness of the −3 part?

    g. What is the linear tolerance?

    h. How many parts are detailed in this drawing?

    i. How many −1 parts are used in the 536A12 drawing?

    j. How large are the holes in the −2 part?

9-6 Refer to Figure 9-17 and answer the following questions:

    a. What is the drawing number?

    b. What is the name of the part?

    c. What is the drawing scale?

    d. How many total parts are there in this assembly?

    e. What assembly is this part used in?

    f. What is the reference number for the 5114 part?

    g. Which parts are held together by the screw?

    h. How is the vice plate fastened to the vice screws?

    i. When was the drawing released?

    j. Who checked the drawing?

9-7 Refer to Figure 9-18 and answer the following questions:

    a. What is the drawing number?

    b. What is the name of the part?

    c. What is the drawing scale?

    d. How many different parts are there in this assembly?

    e. What assembly is this part used in?

    f. What company produced this drawing?

    g. How are parts HC–257 and HC–293 fastened together?

    h. How are parts HC–257 and HC–273 fastened together?

    i. What is the part number for the nuts?

    j. What is the name of part number HC–252?

9-8   Refer to Figure 9-19 and answer the following questions:

   a. What is the drawing number?

   b. What is the name of the part?

   c. What is the drawing scale?

   d. What material is the angle made from?

   e. What assembly is this drawing used in?

   f. What is the length of part 62W15?

   g. What is the linear tolerance?

   h. When are the .378-diameter holes to be drilled?

   i. What is the overall length of the assembly?

   j. What is the latest revision?

9-9   Refer to Figure 9-20 and answer the following questions:

   a. What is the drawing number?

   b. What is the name of the part?

   c. What is the drawing scale?

   d. What is the distance from 65–16209 to 66–4591?

   e. On which side of 66–4591 is 66–4592?

   f. How far from the centerline is 66–4591 located?

   g. How many BM8 rivets are used?

   h. What sheet is this and how many are there in this drawing?

   i. In what zone is the DX8 rivet shown?

   j. What fasteners are used to hold 66–4591 and 66–4592 together?

9-10   Refer to Figure 9-22 and answer the following questions:

   a. What is the drawing number?

   b. What is the name of the part?

   c. What is the drawing scale?

   d. What type of weld is shown in the left side view?

   e. How many parts are there in this assembly?

   f. Who drew this drawing?

   g. How many sheets are there in this drawing?

   h. What kind of weld fastens 101–27–1 and 101–27–4 together?

   i. What company produced this drawing?

   j. What change was made to create revision A?

GLOSSARY

Accuracy   The closeness of a measurement to the true dimension.

Allowance   The tightest possible fit between two mating parts.

Alphanumeric characters   The letters of the alphabet (*A* through *Z*) and the numerals (0 through 9).

ANSI   American National Standards Institute. An organization that publishes engineering and manufacturing standards in the United States.

Apex   The topmost corner of a cone or pyramid.

Arc   A portion of a circle.

Assembly drawing (ASSY)   A working drawing of two or more parts. Shows how they fit together and are fastened.

Auxiliary view   A view projected from one of the six standard views in a multiview drawing. Used to show information not shown in a standard view, such as the true size and shape of inclined and oblique surfaces.

Axis   A straight reference line indicating symmetry or center of rotation for an object.

Baseline   A line used as a common starting point for multiple dimensions.

Basic dimension   A theoretically exact size used as the basis for determining permissible variations.

Basic size   The theoretically exact size of a feature. Tolerances are applied to the basic size to determine the dimensional limits.

Bend angle   The angle at which sheet metal is bent. Measured from the flat condition.

Bend radius   The inside radius specified for a sheet metal bend.

Bisect   To divide in half a line, angle, or the like.

Blind hole   A hole drilled only partway through an object.

Bore   To enlarge a hole by using a boring bar on a lathe or to make a hole smooth, round, or cylindrical by using a boring mill.

Boss   A circular projection, which is raised above a principal surface of a casting or forging.

Box construction   A method for constructing a drawing. The drafter starts by drawing the general framework of an object and completes the drawing by filling in the detail.

Bushing   A removable cylinder sleeve, which is used to provide a bearing surface.

CAD   See *computer-aided drafting*.

Cast   To form an object by pouring molten material into a mold and allowing it to cool.

Centerline (℄)   A line consisting of alternating long and short dashes. Used to indicate symmetry or feature locations.

Chamfer   To bevel an external edge or corner.

Chord   Any straight line that intersects the circumference of a circle at two points. The diameter is a special type of chord.

Circle   A closed plane curved line that is everywhere the same distance from a unique point called the center.

Circumference   The curved line defining a circle or the linear distance around that line.

Computer-aided drafting (CAD)   Drafting a technical drawing by inputting data into a computer and outputting hard copy on a plotter.

Construction line   Light, temporary solid line used for projection and initial drawing layout.

Counterbore   To enlarge the end of a cylindrical hole to a specified depth.

Countersink   To form a conical enlargement at the end of a cylindrical hole.

Cross-hatching   Lines used in a section view to indicate material that has been theoretically cut. Also called section lining.

Curved line   A line that constantly changes direction.

Datum   A theoretically exact feature of an object. Used as an origin for locating other features of the object.

Design drawing   An original drawing showing a design concept. It becomes the basis for the working drawings.

Detail drawing (DET)   A working drawing that shows all the information needed to fabricate a single part.

Diameter (D, Dia. or ⌀)   The length of a straight line (chord) that touches two points on a circle and passes through the center.

Diazo process   A printing process widely used in industry for reproducing copies of technical drawings.

Die   A metal block used for forming or stamping operations or a thread-cutting tool for producing external threads.

Dimensions   Numerical information given on a drawing to indicate the sizes and locations of an object's features.

Double curved   A surface curving in two different directions. Straight lines cannot be drawn on such a surface.

Drafting machine   A complex drafting tool used to draw straight lines that are parallel, perpendicular, and at any angle.

Drafting media   Paper, cloth, or film material on which drawings are created.

Draw   To form metal by a distorting or stretching process, usually by pulling through a die.

Drill   To form a cylindrical hole in an object.

Dual dimensioning   A method of dimensioning that shows both metric (ISO) and inch units for each measurement.

Ellipse   A closed plane curved line. The sum of the distances from every point on the ellipse to two points (foci) on the major (long) axis is constant.

Face   To machine on a lathe a flat face, which is perpendicular to the axis of rotation of the object.

Feature   Any identifiable geometric attribute of an object, such as a surface, hole, or notch.

Fillet   An internal radius of material at the junction of two surfaces joined at an angle.

Fit   The tightness of adjustment between the contacting surfaces of mating parts.

Flange   The top and bottom member of a beam or a projecting rim. Added on the end of a pipe or fitting to make a connection.

Flat pattern   A drawing showing the shape of an object made from flat, thin material prior to forming. The drawing gives the information required for bending the object.

Flat washer   A flat ring-shaped part used under the head of a fastener or nut.

Forge   To shape hot metals through hammering by hand or machine.

Grind   To finish a surface through the action of a revolving abrasive wheel.

Hard copy   A drawing produced by a plotter or printer driven by a computer.

Horizontal   Lines and planes parallel to the earth's surface. At right angles to vertical lines and planes. Level.

Inclined   Lines and planes that are neither parallel nor perpendicular to the earth's surface or the line of sight for a particular view. Slanted.

Installation drawing   A drawing used to show how parts or assemblies are mounted in a larger structure.

Intersecting   The condition where two lines or two planes pass through each other.

Irregular curve   A curved line which is not generally definable.

ISO   The International Organization for Standardization.

Isometric drawing   A type of pictorial drawing widely used in technical drawing. Categorized in the general class of drawing called axonometric.

Key   A part used between a shaft and a hub to prevent the movement of one relative to the other.

Keyway   A longitudinal groove cut in a shaft or a hub to receive a key. A key rests in a keyseat and slides in a keyway.

Least count   The value or distance between any two closest marks on a scale.

Least material condition (LMC)   The condition where a feature contains the minimum amount of material within its size limits.

Limits   The largest and smallest permissible dimensions for a feature.

Lug   A projection of a part. Provides support to allow the attachment of another part.

Mainframe   A computer with a large capacity. Usually supports multiple terminals.

Major and minor axes   Two mutually perpendicular centerlines of unequal length that partially define an ellipse.

Maximum material condition (MMC)   The condition where a feature contains the maximum amount of material within its size limits.

Meter   The unit of distance on which metric linear dimensions are based. A meter (m) is equal to approximately 39.37 in. A meter is divided into 100 units called centimeters (cm), and each centimeter is divided into 10 units called millimeters (mm).

Microcomputer   A self-contained personal computer.

Microprocessor   The central processing unit (CPU) of a microcomputer.

Mill   To machine a piece on a milling machine by means of a rotating toothed cutter.

Minicomputer   A computer system with capabilities between those of a microcomputer and a mainframe.

Monitor   The computer component that displays images or drawings. Also called CRT, graphic display, and screen.

Multiview drawing   Usually, a technical drawing that shows objects by means of more than one flat two-dimensional view.

Mylar   A trade name for polyester drafting film. It is a dimensionally stable media widely used in industry.

Normal   Lines or planes (surfaces) perpendicular to the line of sight for a particular view.

Object line   An unbroken line used to show the features – surface edges, surface intersections, and curved surface limits – of an object.

Orthographic projection   A graphical system in which the object is viewed from an infinite distance and in which the lines of sight are parallel to each other and perpendicular to the projection plane. Both multiviews and isometrics are orthographic projection drawings.

Out of tolerance   The condition where a dimension is outside the stated limits.

Output   Data produced by a computer system.

Pad   A surface projecting slightly from a larger base surface.

Parallel   Lines or planes that are everywhere the same distance apart.

Perpendicular   Lines or planes that are at a right (90°) angle to each other.

Perspective drawing A type of pictorial drawing widely used in art and architecture. The two primary types of perspective drawing are angular and parallel.

Pictorial drawing A drawing that attempts to show objects in three dimensions in a single view. Includes isometric, oblique, and perspective drawings.

Pitch The distance between individual repeated features such as screw threads.

Plane surface A flat surface on which straight lines can be drawn in any direction.

Plan view In a multiview drawing a particular view with a vertical (downward) line of sight. The top view.

Plotter A device used with a CAD system for outputting the hard copy of drawings.

Point A single location in space. A point has no dimensions.

Polar coordinates A system used to locate points on a plane by means of a radius from a given origin and a counterclockwise angle from the positive x-axis, with the origin as vertex of the angle.

Polyester film A dimensionally stable drafting media widely used in industry. Commonly known by the trade name Mylar.

Principal view The view in a multiview drawing that shows the most information. Usually a front or top view.

Projection line A construction line drawn between adjacent views in a multiview drawing. Ensures that the views are in alignment and facilitates the construction of the views.

Projection plane A theoretical drawing surface on which the image of an object is projected. Each view in a multiview has a separate projection plane. Also called image plane, viewing or view plane, and plane of projection.

Punch To perforate a thin piece of metal by shearing out a hole with a nonrotating tool under pressure.

Radius (R) The distance from the center of a circle or arc to the edge of the figure. Equal to one-half the diameter.

Ream To finish a hole to an exact size by using a rotating fluted cutting tool.

Rectangular coordinates A system used to locate points on a plane by means of linear distances along an x-axis (horizontal) and a y-axis (vertical) with a common origin. May also be used to locate points in three-dimensional space by including a z-axis, which is perpendicular to both the x- and y-axis.

Reference dimension A duplicate dimension intended for informational purposes only.

Removed view On a multiview drawing, a view not drawn in its normal projected position. Used when there is not enough room on the drawing for normal projection.

Rib A thin feature whose purpose is to support or strengthen a part.

Round A rounded external corner.

Runout The condition where two surfaces (one curved) become tangent to each other.

Scale (1) A tool used to measure lengths or linear distances. (2) The size of a drawing relative to the actual size of the object being drawn. Drawings may be the same size as the object (full-scale), smaller than the object (reduced scale) or larger than the object (enlarged scale).

Section view On a multiview drawing, a view showing the interior detail of an object with solid lines rather than hidden lines. Section views are drawn as if the object had been cut or broken apart. Used to clarify drawings of complex objects.

Shear To cut off sheet or bar metal through the shearing action of two blades.

Shim A thin plate inserted between two surfaces for the purpose of adjustment.

Single curved surface A surface curved in one direction only. Straight lines may be drawn in only one direction on such a surface.

Slot A narrow cutout feature in a part.

Spoke A rod, finger, or brace that connects the hub and rim of a wheel.

Spotface To finish a round spot on the rough surface concentric with a drilled hole.

Spotweld To weld two overlapping metal sheets in spots by means of the heat generated by resistance to an electric current.

Straight line A line that continues in the same direction for its entire length.

Subassembly drawing A drawing of a part that would be difficult to show on a single drawing of a large, complex assembly. Subassemblies are combined on an overall asembly drawing.

Surface texture The condition (roughness, smoothness, pattern) of an object's surface. Indicated on a drawing by finish marks.

Symmetrical An object that is either exactly the same or a mirror image on both sides of a centerline.

Tangent The condition where a straight or curved line touches at a single point, but does not pass through, a curved line.

Tap To cut an internal thread by screwing into the hole a fluted tapered tool having thread-cutting edges.

Taper To make gradually smaller toward one end.

Template A pattern cut to a desired shape. Used in layout work to establish lines, to locate holes, and so on.

Terminal A computer monitor and keyboard connected to a mainframe or minicomputer.

Tolerance The total amount of permissible variation in a dimension. Tolerance is the difference between the maximum and minimum limits.

Tracing cloth A type of drawing media made of treated linen cloth.

True position The theoretically exact location of a feature specified by basic dimensions.

Truncate To cut off a geometric solid, such as a cone or pyramid, at an angle oblique to the axis.

Units Basic divisions on a scale. The most commonly used units in technical drawing are millimeters, meters, inches, and feet.

Vellum A high-quality paper widely used as a drawing medium. Not dimensionally stable.

Vertex The point where two or more lines or planes intersect, forming an angle(s) between them.

Vertical A line or plane perpendicular to the earth's surface and at a right angle to horizontal.

Web An interior feature or piece of plate or sheet that connects heavier sections of a part.

Weld To join two pieces of metal by heating to the fusion point.

Working drawing A detail, assembly, or installation drawing showing all the information necessary for the production of parts and assemblies.

APPENDIX

Table A-1 Conversion chart—inches to millimeters.

Inches		Millimeters	Inches		Millimeters
Fraction	Decimal	(approx.)	Fraction	Decimal	(approx.)
1/64	.015625	0.397	33/64	.515625	13.097
1/32	.03125	0.794	17/32	.53125	13.494
3/64	.046875	1.191	35/64	.546875	13.891
1/16	.0625	1.588	9/16	.5625	14.288
5/64	.078125	1.984	37/64	.578125	14.684
3/32	.09375	2.381	19/32	.59375	15.081
7/64	.109375	2.778	39/64	.609375	15.478
1/8	.125	3.175	5/8	.625	15.875
9/64	.140625	3.572	41/64	.640625	16.272
5/32	.15625	3.969	21/32	.65625	16.669
11/64	.171875	4.366	43/64	.671875	17.066
3/16	.1875	4.763	11/16	.6875	17.463
13/64	.203125	5.159	45/64	.703125	17.859
7/32	.21875	5.556	23/32	.71875	18.256
15/64	.234375	5.953	47/64	.734375	18.653
1/4	.250	6.350	3/4	.750	19.050
17/64	.265625	6.747	49/64	.765625	19.447
9/32	.28125	7.144	25/32	.78125	19.844
19/64	.296875	7.541	51/64	.796875	20.241
5/16	.3125	7.938	13/16	.8125	20.638
21/64	.328125	8.334	53/64	.828125	21.034
11/32	.34375	8.731	27/32	.84375	21.431
23/64	.359375	9.128	55/64	.859375	21.828
3/8	.375	9.525	7/8	.875	22.225
25/64	.390625	9.922	57/64	.890625	22.622
13/32	.40625	10.319	29/32	.90625	23.019
27/64	.421875	10.716	59/64	.921875	23.416
7/16	.4375	11.113	15/16	.9375	23.813
29/64	.453125	11.509	61/64	.953125	24.209
15/32	.46875	11.906	31/32	.96875	24.606
31/64	.484375	12.303	63/64	.984375	25.003
1/2	.500	12.700	1	1.000	25.400

Table A-2   Conversion chart—millimeters to inches

Millimeters	Inches (approx.)	Millimeters	Inches (approx.)	Millimeters	Inches (approx.)
1	0.0394	36	1.4173	71	2.7953
2	0.0787	37	1.4567	72	2.8346
3	0.1181	38	1.4961	73	2.8740
4	0.1575	39	1.5354	74	2.9134
5	0.1969	40	1.5748	75	2.9528
6	0.2362	41	1.6142	76	2.9921
7	0.2756	42	1.6535	77	3.0315
8	0.3150	43	1.6929	78	3.0709
9	0.3543	44	1.7323	79	3.1102
10	0.3937	45	1.7717	80	3.1496
11	0.4331	46	1.8110	81	3.1890
12	0.4724	47	1.8504	82	3.2283
13	0.5118	48	1.8898	83	3.2677
14	0.5512	49	1.9291	84	3.3071
15	0.5906	50	1.9685	85	3.3465
16	0.6299	51	2.0079	86	3.3858
17	0.6693	52	2.0472	87	3.4252
18	0.7087	53	2.0866	88	3.4646
19	0.7480	54	2.1260	89	3.5039
20	0 7874	55	2.1654	90	3.5433
21	0.8268	56	2.2047	91	3.5827
22	0.8661	57	2.2441	92	3.6220
23	0.9055	58	2.2835	93	3.6614
24	0.9449	59	2.3228	94	3.7008
25	0.9843	60	2.3622	95	3.7402
26	1.0236	61	2.4016	96	3.7795
27	1.0630	62	2.4409	97	3.8189
28	1.1024	63	2.4803	98	3.8583
29	1.1417	64	2.5197	99	3.8976
30	1.1811	65	2.5591	100	3.9370
31	1.2205	66	2.5984		
32	1.2598	67	2.6378		
33	1.2992	68	2.6772		
34	1.3386	69	2.7165		
35	1.3780	70	2.7559		

Table A-3 Standard abbreviations

Adjust	ADJ	Diagonal	DIAG
Advance	ADV	Diameter	DIA
Allowance	ALLOW	Diametral pitch	DP
Alloy	ALY	Dimension	DIM
Alternate	ALT	Double	DBL
Aluminum	AL	Down	DN
American wire gauge	AWG	Draft	DFT
Amount	AMT	Drawing	DWG
Anneal	ANL	Drawn	DR
Approved	APPD	Drill, drill rod	Dr
Approximate	APPROX		
Assemble	ASSEM	Each	EA
Assembly	ASSY	Eccentric	ECC
Automatic	AUTO	Engineer	ENGR
Auxiliary	AUX	Engineering	ENGRG
Average	AVG	Equal	EQ
		Equipment	EQUIP
Baseline	BL	Equivalent	EQUIV
Bearing	BRG	Estimate	EST
Bevel	BEV	Exterior, external	EXT
Bill of materials	B/M		
Blueprint	BP or B/P	Fabricate	FAB
Bolt circle	BC	Fahrenheit	F
Bottom	BOT	Figure	FIG
Bracket	BRKT	Fillet, fillister	FIL
Brass	BRS	Finish	FIN
Brinell hardness no.	BHN	Finish all over	FAO
Broach	BRO	Fitting	FTG
Bronze	BRZ	Flange	FLG
Brown & Sharpe (gauge)	B&S	Foot	FT
		Forged steel	FST
Capacity	CAP	Forging	FORG
Casting	CSTG	Front	FR
Cast iron	CI		
Cast steel	CS	Gage, gauge	GA
Celsius	C	Gallon	GAL
Center	CTR	Galvanized	GALV
Centerline	CL	Galvanized iron	GI
Center to center	C to C	Gasket	GSKT
Centimeter	CM	Grind, ground	GRD
Chamfer	CHAM		
Change	CHG	Half-round	1/2RH
Circle, circular	CIR	Harden	HDN
Circular pitch	CP	Head	HD
Circumference	CIRC	Headless	HDLS
Clockwise	CW	Heat-treat	HT TR
Cold-drawn steel	CDS	Hexagon	HEX
Cold-rolled steel	CRS	High-speed	HS
Combination	COMB	Horizontal	HOR
Concentric	CONC	Horsepower	HP
Copper	COP	Hot-rolled steel	HRS
Counterbore	CBORE	Hour	HR
Counterclockwise	CCW		
Countersink	CSK	Identification	IDENT
Cubic	CU	Illustrate	ILLUS
Cylinder	CYL	Inch	IN
		Included	INCL
Datum	DAT	Increase	INCR
Decimal	DEC	Information	INFO
Deep	DP	Inside diameter	ID
Degree	DEG	Inside radius	IR
Detail	DET	Installation	INSTL
Deviation, develop	DEV	Interior, internal	INT
		Intersect	INT

Table A-3 Standard abbreviations (continued)

Joint	JT	Ream	RM
Junction	JCT	Received	RECD
		Rectangle	RECT
Key	K	Reference	REF
Keyseat	KST	Regular	REG
Keyway	KWY	Reinforce	REINF
		Release, relief	REL
Lateral	LAT	Required	REQD
Left-hand	LH	Reverse, revolution	REV
Length, long	LG	Revolutions per minute	RPM
Letter	LTR	Right-hand	RH
Light	LT	Rough	RGH
Linear	LIN	Round	RD
Locate	LOC		
		Schematic	SCHEM
Machine	MACH	Screw	SCR
Machine steel	MS	Second	SEC
Major	MAJ	Section	SECT
Malleable iron	MI	Sheet	SH
Manufacture	MFR	Specification	SPEC
Manufacturing	MFG	Spotface	SF or SFACE
Material	MATL	Square	SQ
Maximum	MAX	Stainless steel	SST
Measure	MEAS	Standard	STD
Mechanical, mechanism	MECH	Steel	STL
Medium	MED	Stock	STK
Miles	MI	Straight	STR
Miles per hour	MPH	Substitute	SUB
Millimeter	MM	Support	SUP
Minimum	MIN	Symmetrical	SYM
Minute	MIN	System	SYS
Miscellaneous	MISC		
Mold line	ML	Tabulate	TAB
Multiple	MULT	Tangent	TAN
		Technical	TECH
Negative	NEG	Teeth	T
Nominal	NOM	Template	TEMP
Not to scale	NTS	Thick	THK
Number	NO.	Thousand	M
		Thread	THD
Obsolete	OBS	Threads per inch	TPI
Octagon	OCT	Tolerance	TOL
Opposite	OPP	Tool steel	TS
Outside diameter	OD	Total	TOT
		Total indicator reading	TIR
Part, point	PT	Traced	TR
Parting line (castings)	PL	Typical	TYP
Pattern	PATT		
Perpendicular	PERP	Variable	VAR
Piece, pitch circle	PC	Versus	VS
Pitch	P	Vertical	VERT
Pitch diameter	PD	Volume	VOL
Position, positive	POS		
Pound	LB	Watt, wide, width	W
Pounds per square inch	PSI	Weight	WT
Pounds per square foot	PSF	Woodruff	WDF
Product, production	PROD	Wrought iron	WI
Quality	QUAL	Yard	YD
Quantity	QTY	Yield point	YP
Quarter	QTR	Yield strength	YS
Radial	RAD		
Radius	R		

Table A-4   Section line symbols

Shown below are some of the most commonly used section line symbols representing different kinds of materials. A complete set of symbols can be found in ANSI 414.3.

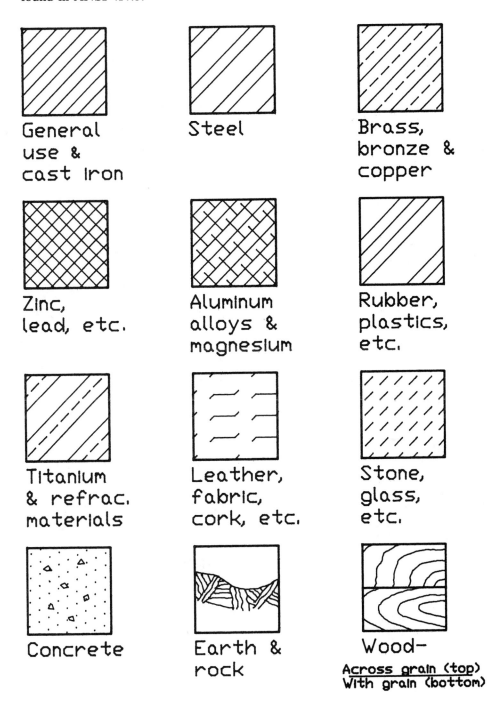

General use & cast Iron

Steel

Brass, bronze & copper

Zinc, lead, etc.

Aluminum alloys & magnesium

Rubber, plastics, etc.

Titanium & refrac. materials

Leather, fabric, cork, etc.

Stone, glass, etc.

Concrete

Earth & rock

Wood— Across grain (top) With grain (bottom)

Table A-5   Electronic symbols

Shown below are some of the most commonly used electronic schematic symbols and their reference designations. A complete set of symbols can be found in ANSI Y32.2.

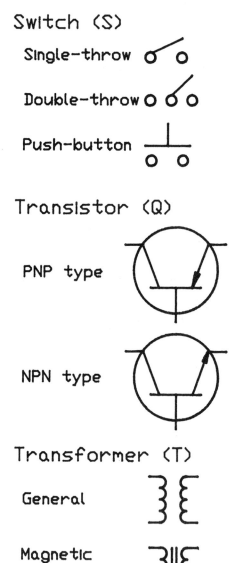

Battery (BT)

Single cell

Multicell

Capacitor (C)

General

Polarized

Variable

Ground

Chassis bond

Frame connection

Rectifier (CR)

Diode Semiconductor

Resistor (R)

General

Variable

Adjustable

Switch (S)

Single-throw

Double-throw

Push-button

Transistor (Q)

PNP type

NPN type

Transformer (T)

General

Magnetic core

Table A-6   Fluid-power symbols

Shown below are some of the most commonly used fluid-power schematic
symbols. A complete set of symbols can be found in ANSI Y32.10.

Pressure control valve

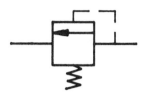

Cylinder

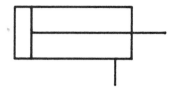

Flow control valve

Fluid motor

Directional control valves

2-Way, 2-Position manually actuated

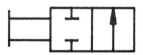

Fluid pump

4-Way, 3-Position solenoid actuated

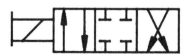

Check valve

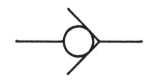

Reservoir & filter

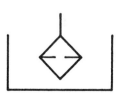

Table A-7   Pipe-fitting symbols

Shown below are some of the most commonly used pipe-fitting and plumbing schematic symbols. Pipes are normally shown as single lines; symbols are used wherever pipes or fittings join. There are five basic types of joints:

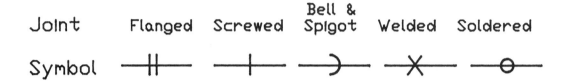

The same fitting symbols are used for all types of joints—in most cases, with only the joint symbol differing. The examples shown are for screwed joints. A complete set of symbols can be found in ANSI Y32.4.

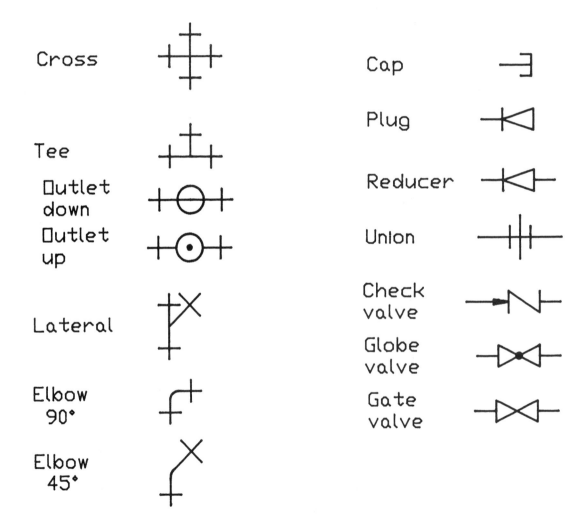

Table A-8 Welding symbols

Shown below are some of the most commonly used welding symbols. A complete set of symbols can be found in ANSI Y32.3.

Basic weld symbols
All symbols are attached to leaders as shown.

 The symbol above the line indicates weld near or arrow side only.

 The symbol below the line indicates weld far or other side only.

 Symbols on both sides of the line indicate weld on both sides.

 A circle at the leader break indicates weld all around the joint.

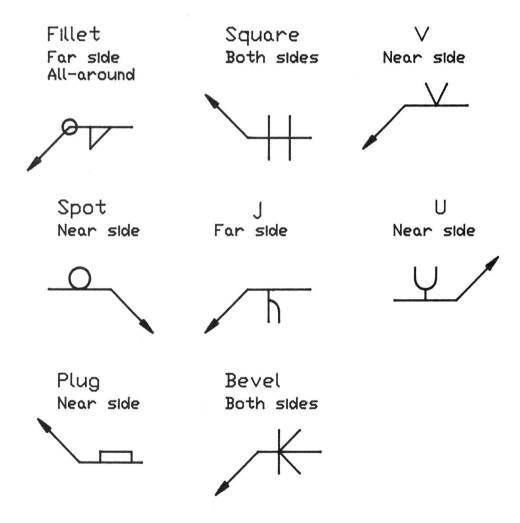

Fillet
Far side
All-around

Square
Both sides

V
Near side

Spot
Near side

J
Far side

U
Near side

Plug
Near side

Bevel
Both sides

345